A Fish Caught in Time

A Fish
Caught
in Time

The Search for the Coelacanth

SAMANTHA WEINBERG

HarperCollins*Publishers*

Epigraph on p. vii from *Verses from 1929 On* by Ogden Nash. Copyright © 1968, 1969, 1970, 1971 by Ogden Nash. By permission of Little, Brown and Company, Inc.

HarperCollins books may be purchased for educational, business, or sales promotional use. For information please write: Special Markets Department, HarperCollins Publishers Inc., 10 East 53rd Street, New York, NY 10022.

FIRST EDITION

Designed by Phil Mazzone

Printed on acid-free paper

Library of Congress Cataloging-in-Publication Data

Weinberg, Samantha, 1966–
    A fish caught in time : the search for the coelacanth / Samantha Weinberg. — 1st ed.
      p. cm.
    Includes bibliographical references.
    ISBN 0-06-019495-2 (hc. : alk. paper)
    1. Coelacanth. I. Title
    QL638.L26 W45 2000
    597.3'9—dc21   99-044800

00 01 02 03 04 ❖/RRD 10 9 8 7 6 5 4 3 2 1

To Mark,
hook, line, and sinker

Consider now the Coelacanth,
Our only living fossil,
Persistent as the amaranth,
And status quo apostle.
It jeers at fish unfossilized
As intellectual snobs elite;
Old Coelacanth, so unrevised
It doesn't know it's obsolete.

Ogden Nash

CONTENTS

ILLUSTRATIONS

ACKNOWLEDGMENTS

Over the last few years, I have immersed myself in the coelacanth and its special world, but I could never have presumed to write a book about this, most studied of creatures, without the generous help and patience of coelacanth experts and junkies the world over.

There are several people without whom it literally would not have been possible. Robin Stobbs, latterly of the J.L.B. Smith Institute in Grahamstown, opened up his archives and his incredible personal store of knowledge and wisdom to me. He was a constant support and guide throughout the process, and remains a valued friend and e-mail buddy. To him I owe unlimited thanks. Marjorie Courtenay-Latimer endured thousands of my questions over several visits. Without her, the coelacanth would probably still not be known to the world, and I could certainly never have written about it. She is a truly special person, an inspiration to us all. Hans Fricke is an extraordinary adventurer, and knows more about the living coelacanth than anyone else—and he could not have been more generous in inviting me to see his submersible, *Geo*, in Germany, recounting his wonderful undersea voyages, and correcting my shaky science. Jerry Hamlin, creator of the dinofish.com website, is another true believer, devoted to saving the coelacanth. He gave me hours of his time. Thank you. I descended on the peaceful island paradise of Mark Erdmann and Arnaz Mehta with little notice, and for nine weeks lived next door and shared their boat, their dinners, and their adventures. For eight months, while the discovery of the Indonesian population of coelacanths was still a secret, they trustingly kept me in the picture, and to them and to their team—Daeng Said, Tantes Ita, and Meli—I dedicate the happy ending of this book.

The research process took a year, and encompassed four continents and numerous people. In the United States, thank you to Roy Caldwell and Keenan Smart, who shared the secret of the Indonesian coelacanth, and to Susan Jewett; to Eugene Balon in Canada; in England, to Anthony Gardner, who pushed me off in the right direction, to Henry van Moyland for his inspired title, to Peter Forey of the Natural History Museum, and to Quentin Keynes, who visited the Comoros soon after the 1952 coelacanth adventure and kindly showed me his treasures; in Germany, to Karen Hissmann and Jürgen Schauer; in South Africa, to Jean Pote, Phil Heemstra, Paul Skelton, and the entire staff of the J.L.B. Smith Institute of Ichthyology; to Bob and Gerd Smith, and William Smith, who shared his memories of his illustrious and unorthodox parents and who generously gave me permission to quote from his father's book, *Old Fourlegs*; to Philip Tobias, Gill Vernon of the East London Museum, and Mike Bruton. Many old and new friends in the Comoros were a tremendous aid: Papa Claude, Christian Antoine, and the staff of Le Galawa Beach hotel, who rescued us from deportation, Ali Toihir, Mouzaoir Abdallah, Mahmoud Aboud, the staff of CNDRS. A special thanks to Said Ahamada, who continues to work tirelessly for the Comoran coelacanth, and whose entire family welcomed me so warmly, and allowed me a glimpse into the world of the night fishermen; and to all the Comoran fishermen, who answered my questions with patience and humor. In Indonesia, thank you to Om Lameh Sonathon and his family, the most gracious of hosts; and to Om Maxon Haniko and their teams of fishermen, and Michael and Corrie and the ladies of MC.

Thank you to Gillon Aitken, agent and friend, to my editor, Virginia Bonham-Carter, who has been with me every inch of the way; to James Kellow and the rest of the good folk at Fourth Estate. Thank you to my American editors, Larry Ashmead and Joe Wojak of HarperCollins, for their valuable input. Thank you to my family and my friends, who have been a constant support: my father and my sister, Joanna, the best of literary critics, my sister Kate, cousins Dan and Ann Simon, and my grandmother, Lilian Le Roith, whose car and house we took over during our three months in South Africa, and who remains my most uncritical supporter. Most of all, to Mark Fletcher, to whom this book is dedicated: the best traveling companion, editor, and husband a girl could wish for.

INTRODUCTION

I saw my first coelacanth on February 5, 1992. It was a sultry day in the Comoros, a remote and beautiful archipelago on the western fringe of the Indian Ocean. I was researching a book about a French mercenary, Bob Denard, who had fallen in love with the tiny outpost, and staged successive coups over more than a decade, in a strong-arm attempt to make the islands his own. Only a few weeks into a six-month stay, I too was well on the way to becoming entrapped by the mysterious charm of the Comoros.

After several hours exploring the warren of narrow lanes of the medieval old town of Moroni, the capital, I was hot and bathed in sweat. I took refuge inside the grandly named Centre Nationale des Réchèrches Scientifiques—the country's only museum—hoping to find a cool space to catch my breath. But it was just as hot—only the bank has air-conditioning—and as I was about to leave, a strange exhibit caught my eye: a glass tank containing a large, stuffed, on the face of it ugly fish. It was unlike any fish I had seen before—its body was covered in scaly armor and its fins were attached by fat limb-like protuberances. It had large, yellowy-green eyes, and a surprisingly gentle expression on its prehistoric-looking face.

The sign beside the exhibit identified the creature as a coelacanth, the marine equivalent of a dinosaur—only older—and a present-day inhabitant of Comoran waters. But there was little more information. I studied the coelacanth for a long time, rolling the strange syllables of its name around my tongue. It conjured up distant memories of nature classes at school, and picture books of bizarre, long-lost creatures. How, I wondered, had it survived, virtually unchanged, for all that time, lost at the bottom of the vast, cold ocean, watching silently as other creatures evolved and became extinct? *Homo sapiens* first walked on earth only a hundred thousand years ago; the fish before me, suspended in murky formalin, pre-dated modern man by 399.9 million years.

The coelacanth stayed in my mind during the rest of my time on the Comoro Islands. There was a shabby hotel named after it; I visited a taxidermist on Anjouan—the second-largest island—who had one on display. In my book about Denard, I used the coelacanth to illustrate the remoteness of the islands, and when I got home, I wrote a story using it as a metaphor for innocence, its discovery a metaphor for colonization. Since then, it has stayed in my thoughts, insinuating itself into conversation, trespassing through my dreams.

At the end of 1997, I decided to do something about it. I started looking into the story of the coelacanth's discovery—digging up documents and papers from the bowels of the Natural History Museum in London—and the more I discovered, the more I was hooked. When my idea for a book about this most unlikely of subjects was commissioned, I was overjoyed, and immediately set out for my beloved islands.

I spent several months in the Comoros, over two visits, much of the time with the night fishermen from the south of Grande Comore, the largest of the four islands, and a known habitat of coelacanths. One night, I was taken fishing by Hassani Ahamada, a veteran of many years of lonely nights on the ink-black ocean.

We met on a black, lava rock beach near the village of Itsoundzou, in time to launch our small wooden pirogues in the last, pink light of day. Two to each boat, we pushed them off the rocks, then paddled quickly out to sea. My fellow fishermen were a quiet group: small and wiry, they wore tatty T-shirts and holey shorts, and battered palm frond hats on their heads, even though the sun was below their sightline. A short way from the shore, Hassani stopped briefly to let out a line, and within a minute he brought it in again, with a small silvery fish on the hook, to use as bait.

He then squatted on the front of two narrow benches that cross the narrow boat, knees under his armpits, beckoning to me to do the same. While I struggled to keep upright, Hassani rowed out to sea with strong, deft strokes, wielding his single short paddle first to one side, then to the other. His movements were swift, but controlled—even with its pair of spidery outriggers, the small pirogue is surprisingly unstable, and a sudden movement to one side could easily topple it.

As the sun slunk beneath the horizon, vertical clouds rose out of the sea, low-lit like a ghostly Gotham city. It was extraordinarily beautiful—a sky of prehistoric dimensions. Night comes quickly in the tropics—a heavy curtain drawn across the heavens—and it was soon dark. There were no lights on the shore, save for the occasional car wending its way along the coast road (even today there is no electricity this far south on Grande Comore), yet it was not lonely. I could hear the muted *plunk* of paddles hitting water, and when my eyes grew used to the night, I could see the dark silhouettes of our fellow fishermen.

They seemed to know exactly where to go. After a lifetime of nights on this inky ocean, they are intimately acquainted with every reef and cave, every dip and rise of the ocean bed. About five hundred meters from shore, Hassani carefully laid his paddles across the boat and prepared the line. He tied two flat black stones, collected from the beach, about eighteen inches above the baited hook, then let out the single line until it touched the sea bed, many hundreds of meters below. When the stone sinkers hit the bottom, he raised the line and jerked free the rocks. He then jiggled the bait along the ocean bed with a quick seesawing action of his arms, feeling its movement through his fingers.

I watched quietly, drinking in the silence of the warm night. We were both waiting, Hassani and I, for a quick tug on his line. We stayed out on the ocean for half a night, beneath the star-filled southern skies, adjusting our position from time to time, occasionally hooking a small fish. Although I knew we had less than a sliver of a chance of catching a coelacanth, I was still half hoping, half dreading that we would haul the magnificent, man-sized fish up from the ocean depths. But a coelacanth never so much as flicked a fin in our direction, and I think I was pleased.

And when a barely perceptible lightening of the sky indicted the coming of dawn, we paddled swiftly back to shore, pulled our boat over the smooth, rounded black lava rocks, and carried our catch to the village to wait for the market to open at daybreak.

I could have gone out with the night fishermen again and again: the rhythm of their lives seeps into one's soul. They have been fishing in the same way for hundreds, perhaps thousands of years—like the coelacanth, the ebbs and swells of the ocean are as familiar as heartbeats. I hope, wish, pray that both man and fish will be allowed to continue their quiet lives for a hundred, a thousand, perhaps a million more years.

LATIMERIA
CHALUMNAE

Southern Africa

December in East London is hot and humid. An ochre haze smothers the small South African city; even the ocean breeze does little to dispel the seasonal lethargy. The year is 1938; *Gone With the Wind* is about to open in America, and Hitler is menacing central Europe. But on the southern tip of Africa, three days before Christmas, most people's minds were on the approaching holidays: offices were beginning to close, families were drifting home to put the finishing touches to their festive arrangements.

At the East London Museum, the thoughts of the young curator, Marjorie Courtenay-Latimer, were far from the upcoming festivities. A small woman with unruly dark hair and lively black eyes, she was surrounded by bones, racing to complete the assembly of a rare fossil dinosaur she and a friend had excavated in Tarkastad.

At quarter to ten in the morning, a shrill ringing echoed through the two rooms of the tiny museum, shattering the young woman's concentration: the telephone had been installed only two days previously. Mr. Jackson, manager of the Irvin & Johnson trawler fleet, informed her that Captain Hendrik Goosen had just arrived at the docks. "There is a ton and a half of sharks for you on the trawler *Nerine*," he said. "Are you interested?" Marjorie was tempted to say no. She wanted desperately to complete the fossil display before the museum closed for the holidays, and she already had a load of fish specimens from Captain Goosen's last voyage, waiting to be mounted. "But I thought of how good everyone at Irvin & Johnson had been to me, and it being so near to Christmas, I thought the least I could do would be to go down to the docks to wish them the compliments of the season." She grabbed a grain sack and called her native assistant, Enoch, and together they caught a taxi to the wharf.

"I went in to see Mr. Jackson," she recalls, sixty years later, "and as I was going out, he said, 'Well, I don't think it's quite a ton and a half of sharks, but a Happy Christmas to you!' They used to torment the life out of me." She hitched up her cotton dress and climbed onto the 115-foot *Nerine*. The crew had all gone ashore except for an old Scotsman, who told her that the specimens were on the fo'c'sle deck. She looked at the pile of fish: sharks, seaweed, starfish, sponges, rat-tail fishes, all kinds of things. She told the Scotsman she probably would not be taking anything; nevertheless she sorted them out carefully. It was then that she noticed a blue fin sticking up from beneath the pile.

"I picked away the layers of slime to reveal the most beautiful fish I had ever seen," she recounts. "It was five feet long, a pale, mauvy blue with faint flecks of whitish spots; it had an iridescent silver-blue-green sheen all over. It was covered in hard scales, and it had four limb-like fins and a strange little puppy dog tail. It was such a beautiful fish—more like a big china ornament—but I didn't know what it was."

"Yes, miss, it's a strange one," the old Scotsman said. "I have been trawling for over thirty years, but I have never seen its like. It snapped at the captain's fingers as he looked at it in the trawl net. We thought you would be interested." He told her that it had been trawled at a depth of forty fathoms, off the mouth of the Chalumna River, and that when Captain Goosen first saw it, he thought it so beautiful that he wanted to set it free. Marjorie said she would definitely take this one back to the museum.

She and Enoch eased the large fish—it weighed 127 pounds—into the sack and carried it to the taxi. The driver was horrified. "I refuse to take any stinking fish in my new taxi!" he exclaimed. Marjorie replied: "It is not stinking. It is perfectly fresh, and if that is the case, I will get another taxi. I brought you here to collect fish for the museum." He relented and they carefully lowered the fish into the boot of the car.

"I was confused," she relates. "I had certainly never seen anything like it before, yet there was a voice nagging in my head. I kept on thinking back to school, where I had written lines about a ganoid fish—an ancient group distinguished by their heavy armor of scales. I had a teacher, Sister Camilla, whose father was a paleontologist at Uppsala University in Sweden, and he used to teach his daughter about marine paleontology, and so she was always teaching us about fish. And on this day I wasn't paying attention, and she turned to me and said: 'You-little-Latimer—what's a fossil fish?' And you-little-Latimer didn't know, because she hadn't been listening. 'You-little-Latimer will write twenty-five lines: A ganoid fish is a fossil fish. A ganoid fish is a fossil fish.' And you-little-Latimer wrote it out twenty-five times. I've still got the book. And so, back at the museum, as I stared at the strange scales on this fish, those lines kept going around in my head: A ganoid fish is a fossil fish—in other words, a fish that has long since become extinct and is known only from fossil records. The scales, the four limbs, all pointed towards it being a ganoid fish. I was so near to classifying it as a ganoid fish: but I thought it couldn't be a fossil fish because it was alive. I didn't think it could be. But I just knew it was something valuable."

She searched through K. H. Barnard's *A Monograph on the Marine Fishes of South Africa*, and any other fish books she could lay her hands on, but could find nothing resembling the strange and beautiful specimen lying on the museum table. That it was something unique and primitive, she could easily tell by its curious structure, its head plates and fin formation. She noted the extraordinary absence of blood or any slimy discharge from its mouth, nose, or body. She took measurements and made rough sketches of the fish. At noon, the chairman of the museum, Dr. Bruce-Bays, dropped by, and Marjorie showed him her exciting new fish. "He was a doctor of medicine, a very sarcastic gentleman. He used to call me Mistress Madge. 'Mistress Madge, it's nothing but a rock cod,' he said. 'You're making such a fuss about it, but it's nothing but a rock cod. All your geese are swans.'"

Many people would have given up at this point and consigned the unidentified fish to the pre-Christmas rubbish dump. But Marjorie was convinced she had something special, and was determined to preserve the fish until she could get someone to identify it. Enoch was sent to fetch a small hand cart, which a member of the board lent the museum whenever they had to transport heavy items. When he returned, they heaved the fish onto the cart and started to push their peculiar load into town.

"My first thought was to take it to the mortuary. At that time, the mortuary was right down at the park." Marjorie recounts. "So the two of us walked all the way there. It was very hot and everyone was very annoyed because they had to get out of the way for us. They shouted, saying why couldn't we walk on the street. Anyway, we got down to the hospital and I went to see the man in charge of the mortuary, a tall man by the name of Evans. When I suggested that our fish could go into his mortuary, he drew himself up to his full six feet, and with his eyes almost popping out of his head, he looked down his nose and said, 'What an iniquitous request! Whatever is everybody going to say?' I said, 'Agh, well they're all asleep, and it's such a beautiful fish.' 'No,' he said. No, he couldn't possibly put a fish into the mortuary. Well, that was that."

She next tried the cold storage. "There was a gentleman there by the name of Latimer, no relation of ours. He at least had the decency to come and look at the fish, but he also said no, he wasn't putting any stinking fish into the cold storage. I don't suppose he was wrong because it could have been giving off gases and there was food in the cold storage. So that was that with him too."

At the time, those were the only two refrigeration facilities large enough to accommodate a five-foot, nine-stone fish. Marjorie began to feel despondent: she knew she had to find some way of saving it. "Then I thought of old Mr. Robert Center, the taxidermist. He had known me since I was a little girl, and had taught me taxidermy, and I was sure he would help me somehow. By now, I realized more than ever that I had something very, very strange—those limb-like fins and the scales, all silvery and iridescent. It was so beautiful."

They arrived at Mr. Center's, and Marjorie showed him her fish and explained the difficulties she had experienced in trying to preserve it. She asked him whether he had ever seen a fish like it, and whether he had any idea what it could be. Mr. Center looked at it; by then it was afternoon and the fish's color was fading to a dusky gray. "No," he admitted, he had never seen a fish like that and he had no idea at all what it could be. He told Marjorie to put it on the table in his workroom. "If we could get some formalin, we could wrap it up and perhaps get someone from Rhodes University to identify it," he suggested. Marjorie agreed. "I said, 'Yes, I will get Dr. Smith' (as he was then)."

She went to a chemist friend in search of formalin with which to preserve the fish. It was at that time scarce, and only a few chemists kept stock for the hospital, so she was able to get only a little—about a liter—which she took back to center. "We diluted it, and we got *Daily Dispatch* papers, saturated them in the liquid, and then wrapped the fish up, very carefully. We needed a sheet of some sort. We asked Mrs. Center, but she wasn't parting with a sheet, so I walked all the way home—there were no buses or anything—and I explained to Mommy what was happening and that I needed to wrap the fish in some cloth so it wouldn't go off until I could get in touch with Dr. Smith. She gave me a double bedsheet. We took it and wrapped the fish really beautifully, with all the formalin-soaked *Daily Dispatch* pages inside."

James Leonard Brierley Smith, a chemistry lecturer at Rhodes University and amateur ichthyologist, acted as honorary curator of fishes for the smaller museums along the south coast. Marjorie had first met him five years previously, on the coast, when she was busy collecting shells and unusual seaweeds for the museum. A spry man, almost drowned by his baggy shorts, with startling blue eyes and a thick bristle of sandy hair, had come up to her and asked what she was doing. She had explained that she worked at East London Museum, and so the friendship began. Smith often visited Marjorie at the museum, and helped her to classify unusual fishes. He was not, however, at Rhodes when she tried to telephone him on December 22, 1938. She left a message, but when he hadn't gotten back to her by the next day, she wrote to him, enclosing a rough sketch of the fish:

EAST LONDON MUSEUM
23rd Dec. 1938

Dear Dr. Smith,

I had the most queer looking specimen brought to my notice yesterday, the Capt of The trawler told me about it so I immediately set off to see the specimen which I had removed to our Taxidermist as soon as I could. I however have drawn a very rough sketch and am in hopes that you may be able to assist me in classing it.

It was trawled off the Chulmna coast at about 40 fathoms.

It is coated in heavy scales, almost armour like, the fins resemble limbs and are scaled right up to the fringe of filament. The Spinous dorsal, has tiny white spines down each filament.

Note drawing inked in red.

I would be so pleased if you could let me Know what you think, though I know just how difficult it is from a discription of this kind.

Wishing you all happiness for the season.

Yours sincerely,

M. Courtenay-Latimer

Daily, she waited for his reply. "Nothing happened, nothing happened, nothing happened," she recalls. Christmas passed in a fishy blur. "I began to feel very despondent. I couldn't think about anything but the fish—my family couldn't understand what was the matter, but

EAST LONDON MUSEUM

ALL SPECIMENS AND EXHIBITS FOR
THE MUSEUM TRAVEL FREE BY POST
OR RAIL IF ADDRESSED:
O.H.M.S.
CURATOR. MUSEUM. EAST LONDON.
PHONE 2995.

East London
SOUTH AFRICA.

23 ~Dec. 193 8.

Dear Dr Smith,

    I had the most queer looking specimen brought to notice yesterday, The Capt of The trawler told me about it so I immediately set off to see the specimen which I had removed to our Taxidermist as soon as I could . I however have drawn a very rough sketch and am in hopes that you may be able to assist me in classing it.

    It was trawled off Chulmna coast at about 40 fathoms.

    It is coated in heavy scales, almost armour like., the fins resemble limbs, and are scaled right up to a fringe of filment.

    The Spinous dorsal, has tiny white spines down each filment.

    Note drawing inked in in red.

    I would be so pleased if you could let me Know what you think , though I know just how difficult it is from a discription of this kind.

    Wishing you all happiness for the season.

Yours Sincerely.
M. Courtenay-Latimer

Marjorie Courtenay-Latimer's letter to J.L.B. Smith.

I just knew, deep in my bones, that it was something important." Then Boxing Day: still no word. "I checked the post every day, and waited for a phone call, but there was no word from Smith."

    It was a shimmering hot week. Every afternoon Marjorie used to go down to Mr. Center to check on the fish, and although it all looked intact, by December 27, oil had started seeping out. Center said he was worried that the loss of oil would begin to destroy the fish. Marjorie didn't want to risk it, so she gave him the go-ahead to skin it, only not from the side, as was normal practice in mounting fish in those days, but rather right down underneath the stomach, so as to save all the scales. It was a hard, slow job. Center cut carefully through the thick scales. The flesh below was pure white, and could be worked like clay. It did not appear fibrous, or like any fish flesh either of them had seen before. There were no ribs, and there was only a flexible tube where the spine should have been. As Center cut into it, pale yellow oil just poured out. Marjorie saved a whole bottle of the fine oil, which she set aside for J.L.B. Smith, and the hard, bony tongue, which she took home to examine. She told Mr. Center that if she had not heard from Dr. Smith by the next evening, he could dump the insides and continue with the mounting.

"There was no word. We waited and waited every day for a letter from Dr. Smith. It was terrible, really. The New Year passed, and we were still waiting." It would be an excruciating eleven days before Marjorie eventually heard from J.L.B. Smith.

Although Marjorie Courtenay-Latimer was not a trained ichthyologist, she was certainly knowledgeable in the field. Even before she was born, she was destined to become a naturalist. Her father, Eric Henry Courtenay-Latimer, wrote in his diary, two months before her birth, "Willie [his wife] and I both pray it will be a lover of all that is beautiful in nature. Willie wants it to be a botanist—I want it to be a lover of animals and birds."

Their first daughter, Marjorie Eileen Doris Courtenay-Latimer, was born more than two months prematurely on February 24, 1907 (a Piscean, appropriately). She weighed a mere one and a half pounds and was given little chance of survival. To her parents, she was a thing of wonder, and they recorded every detail of her life. "The tiny morsel looked like a miniature doll wrapped in cotton wool," her father wrote. "It had masses of dark hair, no finger-nails or toe-nails. . . . Our little treasure is the most fascinating baby in East London . . . our joy in [her] seems more so that it is frail and small."

The Courtenay-Latimers were not a wealthy family, and they had a fairly itinerant lifestyle, following Eric Courtenay-Latimer's job in the South African Railways from one station to the next. Yet they were happy, and took enormous pleasure in outdoor life, in picnics and long seaside walks. On her first birthday, they took their still fragile daughter to Cape Argulhas, on the southernmost tip of the African continent. "Margie was thrilled with the beach and fell in love with a special shell which she played with all day, and finally fell asleep hanging on to it," her father recorded. Aged two, she was found in the duck pond—she had collected all her aunt's ducklings in her pinafore and wanted to take them to bed with her.

Despite her healthy lifestyle, she was often ill. On several occasions, it looked as if she was going to die. "She is thin and frail," her father wrote, "but has a very determined little character, a quaint, little, serious-minded child, with a deep affection for animals, birds, flowers, and her mother and sister."

Birds were Marjorie's special interest: she used to spend long hours watching their nests, collecting eggs and feathers, studying their behavior. When she was eleven, she announced that she would write a book on birds someday. She also had a wonderful butterfly collection, and started gathering ferns and old stone implements. She was a great defender of animal rights, once getting into a fight with her cousins when they wanted to drop a kitten down a well. She came into conflict with her parents one day over a lily: "Margie got a whipping and was sent to bed because she argued with her elders about a Lotus Lily," her father wrote. "She insists it is a Water Lily and not a Lotus. When I asked her why, she said, 'Lotus is a dreadful name for a beautiful plant and Water Lily is a beautiful name.'"

Margie excelled at school and came top of her class in everything except mathematics. Her father wished he could afford to send her to boarding school, but Dr. Brownlee, one of the many botanists Margie had befriended, reassured him: "No teaching could be found to teach her what she loves best in life; the beauty of nature is a gift that she has inherited—the knowledge she has gained in her short life is not found within the four walls of a school. Latimer, be patient, this child will go far in her natural gift of the beautiful. God grant her the health." When Margie nearly died of diphtheria, Dr. Brownlee stayed with her every night and nursed her back to health.

At fifteen, she was sent to a convent school, where she first came into contact with Sister Camilla and her fossil fishes. She continued to do well in all subjects, including music: "She has a lovely touch," her father recorded. "In a way she seems to speak her nature in her playing. She has grown a very gentle loving girl, always ready to help others, and is a great help to her mother. Not pretty, but has a refined face with sparkling, mischievous eyes when she laughs."

Today in her early nineties, Marjorie Courtenay-Latimer still has sparkling eyes when she laughs, which is often. She lives in East London, with Cindy, her fox terrier, in a small house next door to the one she used to live in with her family—where she spent the agonizing Christmas of 1938. There are books every-where—mainly about nature—and jars of shells, woven baskets, flowers, and a life-size, not-yet-finished, red clay model of a Xhosa woman. Another unfinished sculpture is covered by cloths; she lifts them to reveal a startling likeness of J.L.B. Smith. She has recently taken up painting flowers on ceramic tiles, which are placed care-fully to dry on the windowsill. Almost seventy years have passed since she started work at the East London Museum, but her time there, first as curator and then as director, was clearly the most important period in her life.

"I always dreamed of working in a museum," she recounts. "Second to that, I wanted to be a nurse." When she was twenty-one, she became engaged to Alfred Hill, a childhood friend. He was a handsome, party-loving boy. "We used to go up and canoo-dle on the hill near my mother's people's farm on Addo Heights. From there we could see the lights of the Bird Island lighthouse sweeping across the sea." Marjorie developed a compulsive desire to visit the tiny, remote isle. A year later, she broke off her engagement. "He didn't like my madness in collecting plants and climbing trees after birds," she recalls with a chuckle. "He said it was a sissy game and no wife of his was going to do that. Then I fell in love with Eric Wilson, whose father owned the steel factory, but he died and I was heartbroken. He was the love of my life and I never fell in love again."

She decided to become a nurse and got a place on a training course in King William's Town. A few weeks before she was to have started, however, she was invited by one of her naturalist friends, Dr. George Rattree, to chase her dream and apply for the job of curator of East London's first museum—then still under construction. "I went to meet the board, which consisted of the mayor and all these old gentlemen. I was shivering with fright. Then Dr. Bruce-Bays [the chairman] asked me if I played the piano. 'Yes,' I said, so low and weak I wondered if he heard me. They asked me all sorts of questions about what I liked doing. Dr. Rattree said, 'Do you know anything about platanna?' Now, a platanna is a frog, and I said, 'Oh yes,' and told him chapter and verse about the breeding of these things and where they could and couldn't be found. Twenty-five girls were interviewed that morning, and they were all beautifully dressed. I was wearing a home-made dress with bluebells of Scotland all over it and a funny little straw hat. I didn't think I would get the billet." Nine days later, she was offered the job at a salary of £2 per month. She was to be in sole charge of setting up the displays of specimens and running the museum. It was her first job. She was twenty-four.

"I took over the museum in August 1931. It was an empty shell. They had a little room downstairs where they had a terrific lot of rubbish. There were six birds, which were riddled with dermestes [a flesh-eating beetle], so I burnt them—it's a wonder they didn't sack me on the spot. They had a bottled piglet with six legs, about twelve pictures of East London which were quite nice, and twelve prints of Kaffir War scenes. That was it. And there was a box of stone implements collected by Dr. Bruce-Bays, which were no more stone implements than my foot was a stone implement. They all went into the dump."

She went home on her first day, consumed with doubts as to how she was going to be able to fill the museum. The next day she returned with an axe and destroyed the "terrible, terrible" display cases that had been built with a donation from a local philanthropist. She gathered together some old evening dresses, china, and jewelery, and stone implements from her own collection, with which, along with her mother's collection of beadwork dating right back to 1858 and her great-aunt Lavinia Walton's Dodo egg, she made the museum display.

The new East London Museum opened on September 23, 1931, and from that day, it became Marjorie's life. "I used to go out on weekends and collect wildflowers, which I would label and try and teach the children their names." She never stopped collecting—every weekend, every holiday, adding large collections of South African sea shells, seaweeds, bird eggs, butterflies, moths, insects, and local ethnological material to the little museum's growing display.

East London Museum earned a good reputation. In 1932, two visiting dignitaries were so impressed by "the lady curator" (as Marjorie refers to herself) that they organized for her to spend six months at the Durban Museum. There she learnt how to mount, model, and classify all manner of specimens. She returned to her museum full of enthusiasm and ambition.

In December 1933, Marjorie met J.L.B. Smith for the first time, and he encouraged her to send him any fish she needed classified. "I was very fond of Professor Smith," she says. "People thought he was a very difficult man, but I always got on with difficult people. He was very exacting and I was scared of getting things wrong in front of him. But I loved and admired old J.L.B. He was the most wonderful person. I was very lucky to have such wonderful friends."

In 1933, she was invited to spend six months at the South African Museum in Cape Town. While there, she met a Mr. Patterson, who was in charge of all the offshore islands, including Bird Island, which had used to distract her attention from kissing Alfred Hill. "I have always believed that I had callings in life: things I just had to do. Bird Island was a calling," she says now. "My mother's people were 1820 settlers and the lights from Bird Island used to flash into one bedroom window of their house on Addo Heights. When I was small it scared me, and my mother would say, 'Don't be frightened, it's just the light guiding the sailors at sea.' I thought what a wonderful place this Bird Island must be. So, when I met Mr. Patterson I had to grab my chance.

"I have never tried to vamp anybody as I did Mr. Patterson. At that time, women weren't allowed to go to Bird Island. But I pestered the life out of him. When I was off duty from the museum, I used to go and visit him, I used to take him fruit and sweets. Finally he said, 'If you get one other woman to go with you, I'll give you a permit to go to Bird Island.' So I came home, and asked my mother to go with me. She said, 'Yes, but what's Daddy going to say?' I said, 'You just tell him we want to go and I've got a permit.'"

"Margie has got round her Mother to go with her to Bird Island," her father wrote. "I am really furious with them. Willie is as bad as Margie. There is no holding them back." But he granted them his permission.

"It was November 1936, and we were in Port Elizabeth, packed up and ready to go," Margie recounts. "The evening before we were due to leave, a wire came from my father saying he was coming to Bird Island with us. But he had no permit. I went to the port captain in PE, howling blue murder, because I thought that after all this time he was going to stop me going. I wanted to get in the tug and go before he arrived, but the port captain said, wait a bit, he would see what he could do. Daddy's train arrived at about 8 A.M., it was pouring with rain, and we had been at the wharf since 5 A.M. with all our own groceries, all our stuff, loaded on the tug. I didn't even want to greet my father. He was very pleased with himself, but I wasn't happy to see him at all, because I still didn't know if they would let him go. However, he got the permit and we all went together.

"We spent about three months there—it was wonderful. Daddy was bored stiff, because it was just when the Prince of Wales was abdicating and there were no newspapers. Of course, Daddy wanted to know all about it. Oh gosh, what a hard time he gave us.

I didn't care, I was on Bird Island and enjoying myself."

"I wonder sometimes if she belongs to me with her strange ideas of pleasure," her father wrote in his diary while he was on Bird Island. "What girl of her age would isolate herself to spend her time at a Godforsaken place like this, yet she is in her glory. She has learnt to shoot and is a crack shot."

Marjorie spent all day on the mile-square island, home to 27,000 birds. She collected, skinned, and observed the breeding habits of gannets, terns, albatrosses with a ten-foot wingspan, and penguins. She collected marine fish, shells, plants, anything that was of interest to the museum. It was cold and she carried a pet rabbit around in her jacket.

She continued to visit Bird Island for years, and it has remained a very special place for her. "I look at my pictures," she says, bringing out a large album of black-and-white photographs, each carefully annotated in faded ink, "and I feel homesick." The pictures show a smiling young girl in a simple cotton frock, amidst thousands of birds. Bird Island was Marjorie Courtenay-Latimer's private heaven, and when it was time to leave, she had collected fifteen packing cases of specimens.

"That was when I met Captain Goosen," recounts Marjorie. "He was captain of the *Nerine*, and he used to stop off at Bird Island, to catch Belgian rabbits for his crew, when they got sick of eating fish. Captain Goosen became interested in my work and very kindly said he would bring back my packing cases to East London, one at a time. When he had finished bringing them in, he said he would like to continue collecting for the museum, so I designed this big tank in which he could keep specimens for the museum and the aquarium. Captain Goosen was charming, a fine man. I was very fond of him. He collected everything—starfish, sharks, all kinds of things. Then he would phone and I would go down to the dock by taxi to bring them back and mount them."

It was Captain Goosen's latest find that was causing the lady curator so much stress. Eleven days had passed since Marjorie Courtenay-

Latimer had sent her letter and sketch to J.L.B. Smith, yet she still had received no response.

The letter, in fact, had been forwarded from Rhodes University in Grahamstown, to Knysna, three hundred and fifty miles along the coast, where J.L.B. Smith and his wife, Margaret, were spending the holidays. Christmas, and then New Year, had delayed its passage. J.L.B. was a slight man, with a weak constitution. When a friend brought around the post on January 3, 1939, he was still weak from a recent illness. Among the letters, he recognized Marjorie's handwriting. He opened the letter and read her description of the fish. He turned to the next page and saw the sketch.

"I stared and stared, at first in puzzlement," he wrote in *Old Fourlegs: The Story of the Coelacanth*, his account of the discovery of the living coelacanth first published in the United Kingdom in 1956. "I did not know any fish of our own, or indeed of any seas like that; it looked more like a lizard. And then a bomb seemed to burst in my brain, and beyond that sketch and the paper of the letter, I was looking at a series of fishy creatures that flashed up as on a screen, fishes no longer here, fishes that had lived in dim past ages gone, and of which only fragmentary remains in rocks are known." He told himself not to be a fool, but the more he looked at the sketch, at the tail, the limb-like fins, and the large scales, the more convinced he became of its close resemblance to a fossil he had seen of a fish thought to have been extinct for seventy million years. "What I suspected was so utterly preposterous that my common sense kept up a steady fire of scorn for my idiocy in even thinking of it," he wrote.

It was a remarkable feat of mental agility. Smith had apparently taken a rough sketch by someone who was not a skilled artist, of a five-foot fish, found in the Indian Ocean off southern Africa, and connected it with a fossil, a little over twelve inches long and 200 million years old, which had been discovered in freshwater in Greenland, and which he had read about in a scientific journal.

Margaret Smith was astounded by his strange behavior. He was standing up, staring at the letter in complete silence. Finally, he turned to her and said, "I know you'll think I'm crazy—but they have found a fish at East London of a type generally thought to have been extinct for many millions of years."

"I *did* think he had a touch of the sun, but nine months of marriage to this brilliant older man had taught me much, and instead of blurting out my incredulity, I asked quietly, 'What makes you make a statement like that?'" she wrote many years later. "You see that tail," he said. "No living fish has a tail like that." She took the letter from him and pointed out the date—it had been written nearly two weeks earlier. J.L.B. immediately started to worry. He knew of the East London Museum's simple facilities, and feared the worst. He sent Marjorie a telegram:

MOST IMPORTANT PRESERVE SKELETON
AND GILLS FISH DESCRIBED

Smith knew he had to see the creature in order to confirm his suspicions, but for some inexplicable reason he didn't immediately set out for East London. In *Old Fourlegs*, he claims he was tied up with marking University of South Africa exam papers and could not responsibly leave the task. He also feared that it might not be the incredible and extraordinary discovery he hoped for. The same day as he received Marjorie's letter, he wrote a longer reply, again urging her to save the fish's soft parts, and confiding his suspicions as to its provenance: "From your drawing and description the fish resembles forms which have been extinct for many a long year, but I am very anxious to see it before committing myself. . . . Meanwhile guard it carefully, and don't risk sending it away." He spent the rest of the day and night, until the post office opened the next day and he could try to telephone her, in a feverish state, with the sketch churning around and around in his brain.

In East London, Marjorie had almost given up hope of hearing from Smith. When eventually she got the wire, urging her to save the viscera, it was too late. Thirteen days had passed since the fish had first come to her, and the taxidermist, Mr. Center, had long destroyed its internal organs and tissues. By mounting it, however, he had managed to save the internal skeleton and the skin, which by this time had been turned brown by the formalin.

Marjorie Courtenay-Latimer's rough sketch
of her unusual "limbed" fish.

When Smith managed to reach Marjorie on the telephone the next day, he stressed the importance of the innards, and sent her off to the municipal dump to try to recover them. Poor Marjorie had no luck: the rubbish of East London was dumped in the ocean.

Smith spent the succeeding weeks in a state of turmoil. The more he read, the more he became convinced that the creature was a Coelacanth (he always thought of it with a capital C), an ancient fish, whose origins date back 400 million years. His confidence was strengthened by a long-held conviction that he was destined to discover "some quite outrageous creature. . . . This was so firmly fixed in my mind that just as my peculiar set of circumstances and qualifications had set the stage ready for the appearance of the Coelacanth, so in one sense had this premonition prepared me to deal with such a fantastic possibility as had now arisen, and, indeed, even while my common sense rejected it, to seek for it in an obviously impressionistic sketch by someone not an ichthyologist."

J.L.B. Smith was experienced enough to know the importance of the discovery: it would be the greatest zoological find of the century. If, however, he announced it as such and was found to be wrong, he would be the laughingstock of the world's scientific community. Until he could get to East London, he would not know for sure. "Those were awful days, and the nights were even worse," he recorded. "I was tortured by doubts and fears. What was the use of that infernal premonition of mine if it was just going to make a scientific fool of myself. Fifty million years!* It was preposterous that Coelacanths had been alive all that time, unknown to modern man."

He wrote to K. H. Barnard of the South African Museum, cautiously expressing his beliefs. The reply was prompt and incredulous. Smith looked again and again at Marjorie's letter and sketch, as if to reassure himself. He wrote to her again:

KNYSNA
9th January 1939

Dear Miss Latimer,

Your fish is occasioning me much worry and sleepless nights. It is most aggravating being so far away. I cannot help but mourn that the soft parts of the fish were not preserved even had they been almost putrid. I am sorry to say that I think their loss represents one of the greatest tragedies of zoology, since I am more than ever convinced on reflection that your fish is a more primitive form than has yet been discovered. It is almost certainly a Crossopterygian allied with forms that flourished in the early Mesozoic or earlier, but which have been extinct for many millions of years. Comparatively little is known of the internal structure of such fishes, naturally nothing of the soft parts, since

---

*It is now believed that the fossil record died out as long as 65 million years ago—at the time of the last Great Extinction.

fossil remains are all that help us to know what they were like. Your fish has the general external features of a Coelacanthid, fishes common in early times in northern Europe and America. Whether or not it is a new genus or family I can determine only on examination, but I feel sure that it will make a great sensation in the Zoological world. . . .

To honour you for having got this wonderful thing I have provisionally christened it (to myself at present) Latimeria chalumnae, *and it may even be a new family.*

Kindest regards,
Yours sincerely,

J.L.B.Smith

"I was terribly upset by that letter," recalls Marjorie. "I was always terrified of doing anything wrong for J.L.B. I phoned and told him that if he had come earlier, nothing would have been thrown away, but I knew it was all my fault, and I have had to suffer that ever since."

Letters flew back and forth between East London and Knysna, unhindered now by inconvenient holidays. J.L.B. wrote again to Barnard in Cape Town, but never asked for assistance. "At no time did I look on it as anything but my own," he explained in *Old Fourlegs.* "There was no question in my mind that I had to take the full responsibility for the decision of the identity of the creature. . . . I had deliberately chosen to carry the terrible responsibility myself, making it indeed my own funeral." He wrote to Miss Latimer, authorizing her to offer the trawler people £20 for another perfect specimen. She sent him some scales, which reinforced his confidence in his classification.

Eventually, on February 16, J.L.B., a slight man who seemed almost to radiate energy, and Margaret Smith, his taller, pregnant wife, arrived in East London. It was pouring with rain, and they went directly to the museum to see Marjorie Courtenay-Latimer

Marjorie Courtenay-Latimer with *Latimeria chalumnae*.

and the coelacanth.* According to her diary, she had walked to work early, and had been at the museum since 6 a.m, in a state of great excitement and agitation. Smith was ushered into the inner room where he saw the fish for the first time, sitting on Marjorie's large mounting table: "Although I had come prepared, that first sight hit me like a white-hot blast and made me feel shaky and queer, my body tingled. I stood as if stricken to stone. Yes, there was not a shadow of doubt, scale by scale, bone by bone, fin by fin, it was a true Coelacanth."

*In his book, Smith wrote that Courtenay-Latimer was out when they arrived. This she refutes vehemently: "That always annoyed me, because I had been so anxious for him to come, and then I think it was put into his book that I was out when Prof. Smith came—as if I was going to go out shopping! I never went shopping; I had little time to shop."

II

EX AFRICA SEMPER
ALIQUID NOVI

Of all the fish that could have possibly, miraculously, "come back to life," the coelacanth was by far the most interesting, and Smith was well aware of its importance. The one-time existence of coelacanths had been known for almost exactly a hundred years by the time Marjorie Courtenay-Latimer came across her beautiful blue fish on December 22, 1938. In 1839, the Swiss scientist, Louis Agassiz, had described the fossil of an unusual fish tail found in a road cutting through the Permian marl slate of Durham in the north of England. He noted that the fin rays supporting the tail were hollow, and as a result, coined the name *Coelacanthus* (from the Greek for hollow spine) *granulatus* (after the tubercular ornamentation on the surface of its scales).

In the succeeding years, a plethora of similar fossils were discovered from excavations around the world: in Germany, England, the United States, China, Brazil, Madagascar, and Greenland. Their defining characteristics were their hollow spines and strange lobed fins, but apart from these, it appeared that coelacanths came in all shapes and sizes. Some were fat, others were thin; they had large bodies and small tails, or large tails and small fins. They ranged in size from a couple of centimeters to a massive three meters. The most ancient specimen, known as *Diplocercides*, was found in Devonian rock, making it between 375 million and 410 million years old. The most recent, of the genus *Macropoma*, only a foot long, was found in freshwater Cretaceous deposits—dating back some 70 million years—in Europe and Asia. No later coelacanth fossils have come to light, and until 1938, it was assumed that the fish, like dinosaurs and the majority of species, had perished in the great extinction at the end of the Cretaceous period.

Using the delicate, intricate fossil records, paleontologists were able to make detailed reconstructions of what they believed the fishes were like, at least in terms of external and skeletal appearance. And the more they found out about the coelacanth fossils, the more excited they became.

This was in part due to a fundamental revolution in biological theory. In 1859, *On the Origin of Species* was published, and in it and his later *Descent of Man*, Charles Darwin told a wondering world that man was created by a process of evolution by natural selection; that our ancestors were monkeys, their ancestors were reptiles, whose ancestors in turn were fish. While gaps remained in the family tree, however, there was still a basis for skepticism. The Christian Church maintained its anti-evolutionary position; the onus to prove the theory of evolution lay with science. Suddenly every creature, past and present, acquired a new fascination, as scientists strove to work out where each of them fit in the evolutionary chain and so uphold Darwin's theory. Bit by bit, they began to piece together the evidence, using fossil remains of extinct animals to support their case. Foremost among these "missing links" was the mechanism by which fish walked out of the sea and began to colonize the land—literally, the biggest step ever taken in the history of evolution. If evidence could be found for a tie between sea dwellers and land dwellers, it would help to shore up the case for evolution, and undermine the creationists.

By the 1930s, when the coelacanth's continued existence came to light, evolution was still a hot topic. (Even as recently as ten years ago, a survey indicated that only half of American adults believe that the theory of evolution has any basis in fact.)

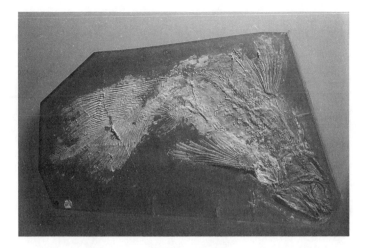

Fossil coelacanth: *Coccoderma suevicum*, 150 million
years old, 32 centimeters long.

Prior to the Devonian period (between 410 and 360 million
years ago), nothing lived on land save for a few spiky, low plants,
some scorpions, and other insects. The earth was congregated into
large continents completely different from the ones we know today,
and in a constant—albeit slow—state of change. There were massive freshwater lakes, and while the land was a bare, desolate place,
these lakes and oceans writhed with life. There were creatures of all
forms and sizes, most of which would be unrecognizable to us
today: small, flat, heavily armored, jawless ostracoderms (bone-
skinned fishes), which dwelled on the ocean floor, their mouths
constantly open in search of food; giant nautilus, the size of a man;
sea scorpions larger than lobsters; the first jawed fishes, the placo-
derms, which were sometimes several meters long; and primitive
giant sharks, even then fearsome predators.

It was during the Devonian period, often called the Age of
Fishes, that the first bony fishes, the vertebrates, appeared on the
scene. Even these would look unfamiliar today, with their encase-
ments of heavy armor, protection from the omnipresent predators.
The vertebrates were divided into two groups: the ray-finned

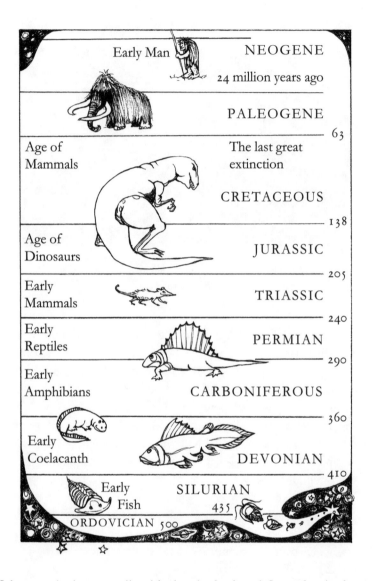

fishes, or Actinopterygii, with the single dorsal fin and paired pectoral and pelvic fins common to most modern fishes; and the lobe-
finned fishes—the coelacanth, the lungfish, and the rhipidistian—
whose fins appeared to sprout from the end of fleshy, limb-like
lobes, almost like toeless legs. These were known as Sarcopterygii

(from the Greek *sarco*, meaning fleshy; *pterygii*, wing or fin), and were characterized also by their extra dorsal fin. Many were predators, and at times, it is clear, they were extremely abundant. Shale from a Triassic swamp uncovered recently during excavations for a new library at Princeton University contained an average of a dozen coelacanth fossils per square foot.

Sometime towards the end of the Devonian period, a single species of freshwater lobe-finned fish evolved legs. In its new guise of Ichthyostega (literally, walking fish) it crawled out of the water to conquer the land—this much scientists agreed upon. What was not so certain was which of the group had evolved into Ichthyostega: the lungfish, rhipidistian, or coelacanth?

Paleontologists subjected their fossil samples to minute scrutiny under ever stronger microscopes. First one, then the next, then the third was pronounced to be our fishy great-grandfather. They pored over coelacanth, lungfish, and rhipidistian fossils for signs of gills or heart or anything that could have preceded those first gulps of air—but without the soft parts in front of them, they were never going to know for sure. The appearance on the scene of a living *Latimeria* opened a fresh chapter in the search for our oldest living ancestor.

In East London, J.L.B. Smith circled the coelacanth several times, in silence. The large fish—turned brown from the formalin—lay on the table in Marjorie Courtenay-Latimer's tiny office. It was a portentous moment. After nearly seven weeks of agony and turmoil, during which he had thought of nothing and dreamed of nothing but the fish, he was finally in its presence. His premonition had been confirmed: he was indeed the first person to recognize an extraordinary and important living creature. He went close to the fish and stroked it. Then he turned to Marjorie and said, "Lass, this discovery will be on the lips of every scientist in the world."

"I was amazed, and surprised and delighted, above all, that at last it had been identified and named," she recalls. "I asked Dr. Smith all kinds of questions, which he answered only vaguely. He

sat down and said, 'To think that a fish of this age could exist!' I
asked, 'How old is it?' He counted the rings on its scales, and said
it was about thirty-three years old, but its origins were more like
70 million years old. My mind boggled. So I had been on the right
track. It was a living fossil."

The night after Smith had identified the coelacanth was a
very restless one, according to his wife. "No sooner did I get to
sleep that he'd waken me and ask, 'Tell me, lass, I'm not dream-
ing, am I?' . . . Or a little later, 'Sorry to waken you—but I'm not
crazy, am I?' "

J.L.B. Smith knew he had to wait for official approval from
his fellow scientists. He planned to announce the discovery of a
new—and living—species of coelacanth in the leading British journal, *Nature,* and at the same time to give the fish its scientific
name. What he didn't suspect was the immediate interest and
enthusiasm it would generate in the lay public. News leaked out
when a journalist and photographer from the *East London Daily
Dispatch* turned up. Marjorie, always conscious of what was best
for her museum, had alerted the paper that the Smiths would be
visiting that day. J.L.B. granted an interview, but initially refused to
let the photographer, Mr. Adams, take a picture. He was worried
that someone else might jump in and name the fish before he had
published his report in a scientific journal. After Marjorie pleaded
with him, he relented, allowing Adams to take one picture, on the
express understanding that it was to be published only in the *Daily
Dispatch.* Perhaps inevitably, Mr. Adams acted in true paparazzo
fashion and sold the picture around the world—even the East London Museum had to pay him two guineas for a print.

The scientific piracy Smith feared, however, never material-
ized. The *Daily Dispatch* interview appeared on February 20,
1939, the same day the coelacanth was put on display at the
museum. From early in the morning, the curious queued around
the block to see the zoological sensation that had been found off
their coast. But while the find had fired the imagination of the public, adding a new word to their lexicon, it still was greeted with a
degree of skepticism by the scientific world. Marjorie received a

phone call from a scientist at the British Museum—at that time God of the natural history world—to ask if she was positive that the fish hadn't been dead, and just capsuled in mud for tens of millions of years before Captain Goosen trawled it up. She replied that she was sure. What proof did she have? "So I said that at 11:30 it was blue in color, but by 5 P.M. it had faded to a dirty gray. They asked again if I was positive, and I said 'Yes, for the last time.'"

Two days later, on February 22, the East London Museum sent the fish by train, with a police guard, to Grahamstown. It was taken to Smith's house and put in a special room. "It had a curious, powerful, and penetrating odor, an odor that in the coming weeks was always to pervade our lives, awake or asleep," Smith wrote. His household was drilled on fish protection measures; it was to be left alone in the house at no time, and in case of a fire, the fish should be the first thing to be saved. Smith's every waking and sleeping moment was consumed by the coelacanth.

After a preliminary examination of the fish, he sent a photograph and report describing the external features of the coelacanth to *Nature* in London. The article began, *"Ex Africa Semper Aliquid Novi*—there is always something new out of Africa (Pliny)." The day it came out, March 18, 1939, Smith's claim to the naming of the fish was sealed. It would forever remain *Latimeria chalumnae* J.L.B. Smith. The species would for always share a name with a pint-sized young woman from East London and an eccentric, obsessive Grahamstown scientist. "When Dr. Smith wrote to say he had named the fish after me," recalls Marjorie Courtenay-Latimer, "I said I thought it should have been named after Captain Goosen, who had brought it for me; without him, there would have been no coelacanth. 'But you ultimately saved it for science,' he said."

The *Nature* article also put an end to most of the scientific skepticism. On March 16, J. R. Norman from the British Museum presented a paper on Smith's article to the Linnean Society in London—the same body to whom Darwin had first revealed his theory of evolution—thus giving it a further official stamp of scientific approval.

Since Darwin coined the term "living fossil" to describe the living "remnants of a once preponderant order," previously known only from fossil records, there had been a costly scramble to find examples of these archaic organisms. Darwin had proposed that living fossils would be found in the ocean depths, which had remained relatively unravaged by the fundamental environmental changes that, he asserted, had been the driving force behind evolutionary change.

It was to explore and survey these depths that a major, three-and-a-half-year expedition was put together by the Royal Society, the British Admiralty, Treasury, and Museum. In December 1872, the H.M.S. *Challenger*, a converted navy warship, set sail from Portsmouth harbor with her crew of 240 sailors and scientists. The once mighty vessel had been stripped of all but two of her guns, and was laden instead with thousands upon thousands of sample bottles, microscopes, and vast vats of pickling alcohol. Among the main aims of the expedition's leader, C. Wyville Thomson, was to bring back examples of living fossils that Darwin had predicted so confidently would live in the depths they were set to explore.

The ship circumnavigated the globe, trawling the oceans and seas from the tropics to the poles, with giant nets and heavier metal dredges, which scooped up soft sediment from the ocean floor. It was by no means a pleasure jaunt. At times the large vessel was buffeted by giant waves that threatened to sweep the decks clean; four sailors and one scientist died, two men went insane, one committed suicide, and sixty-one deserted ship—driven to distraction by the mind-numbing routine of years of dredging.

In spite of all this, the expedition was declared to have been a resounding success. The scientists made great advances in mapping out the topography of the ocean floor, giving a name to a new scientific discipline: oceanography. They confirmed that the depths were teeming with strange forms of life, and identified in all 4,717 new species. They all but failed, however, in their quest for a living fossil. They found only one, a small and not particularly interesting type of squid called *Spirula*.

In the face of the rarity of living fossils, therefore, the discovery of the coelacanth—which was not only a living fossil but one

The East London Fish: The most startling "living fossil" ever discovered.

THE OUTDOOR WORLD

A LIVING FOSSIL CAUGHT IN THE SEA

BEST FISH STORY IN 50,000,000 YEARS

One Of Most Sensational Scientific Discoveries Of The Century

EAST LONDON'S WONDER SPECIMEN TRAWLED

1939 headline from the *Eastern Province Herald*.

believed to be closely related to the lineage of mankind—was a major event. The world's media went into a piranha-like feeding frenzy. From the middle of March 1939, for several weeks, articles appeared in newspapers and magazines from New York to Sri Lanka. The long report in the *Auckland Star* of New Zealand was entitled "Loch Ness Outdone." An almost life-size pull-out picture was published in the *London Illustrated News*, alongside an article by Dr. E. I. White of the British Museum. Headlined, "One of the Most Amazing Events in the Realm of Natural History in the Twentieth Century," it described the discovery as "sensational," and claimed that "this occurrence is as surprising as if one had discovered a living example of the dinosaur *Diplodocus*, the 80 ft reptile of the Mesozoic era." Smith found sections of the article patronizing, however, and "not flattering to remote scientists like myself." White had written, "The report of this discovery was made some time ago, but was treated with justifiable scepticism by the experts—they are only too familiar with deliberate hoaxes or the misplaced enthusiasms of the uninformed to place credence in such reports until supporting evidence is available. (One recalls to mind the instances of the 'Endfield Dinosaur,' actually the remains of a luckless cart-horse, and of the 'Suffolk Mammoth,' the report of which sent an expert racing down into East Anglia to find nothing but the remains of fish-manure arranged in plough furrows.)"

In accordance with Darwin's proposition, White expressed his belief that the coelacanth "almost certainly was a wanderer from deeper parts of the sea to which its kind have retreated in the face of fierce competition with the more active modern types of fishes."

This was an idea that was to become widespread across the scientific community, and one to which Smith reacted with derision: "To me one glance at the Coelacanth disposed of any idea that it lived in the 'inaccessible depths of the ocean'; yet a number of scientists all over the world apparently accepted this with a sigh of uncritical relief. . . . When I looked at that fish, even the first time, it said as plainly as if it could speak: 'Look at my hard, armoured scales. They overlap so that there is a threefold thick layer of them all over my body. Look at my bony head and stout spiny fins. I am so well protected that no rock can hurt me. Of course I live in rocky areas, among reefs, below the actions of the waves and surf, and believe me, I am a tough guy and not afraid of anything in the sea. . . . My blue colour alone surely tells you that I cannot live in the depths. You don't find blue fishes there.'"

A letter was published in *Nature* criticizing J.L.B. for naming the fish after Miss Courtenay-Latimer, since she had done science such a disservice by losing the viscera. To this he responded in the strongest terms: "It was the energy and determination of Miss Latimer which saved so much, and scientific workers have good cause to be grateful. The genus *Latimeria* stands as my tribute," he wrote in *Nature*.

Smith spent every spare hour over the following months back at his house, dissecting the coelacanth and preparing his official monograph for the Royal Society of South Africa. In order to fit in this work around his university duties, he would get up at three in the morning and work on the fish until six. Then he would take a four-mile walk in the hills, write up his observations on his return, have breakfast, and leave for college at eight-thirty. While he was there, his wife Margaret would type up his notes, which he would look over and correct at lunchtime. She would retype in the afternoon, so that he could work again from five until ten in the evening. When his eldest son (from a previous marriage), Bob, got home from school, he would find rough sketches of fish parts with instructions attached to them, waiting for Bob to draw them neatly.

Margaret shared with J.L.B. each step of his dissection: "I witnessed his agony as he found the taxidermist's iron nails or the holes they had made in the head bones. I shared his excitement when he discovered, still attached under the skin, undamaged and undisturbed, the beautiful, delicately shaped extrascapular bones that carry the sensory canal across the nape," she recalled in an article in *Oceanus* magazine in 1970.

It was an intense and stressful period. "We had no social life, business and financial affairs took a back seat, and our food reached its destination over and between sheets of manuscript," J.L.B. Smith wrote. "We had no conversation, no thoughts, no ideas nor eyes, for anything except the Coelacanth, all day and all night. We could never forget it, certainly not with that smell."

It was around that time that the coelacanth earned the sobriquet "missing link," mainly from nontechnical journalists entranced by the story. It had the unpleasant effect of generating a torrent of letters from religious fundamentalists across the world, who were yet to be reconciled to the theory of evolution. They castigated Smith for ignoring the Bible in his "preposterous statements" about millions of years. Did he not know that Adam was created in 4026 BC ? The theory of evolution, they stormed, was evil: an anti-religious invention of the devil put into some men's minds to enable them to divert others from the path of true thought.

Many of these letters, which Smith collected into the "crackpot file," came from South Africa.* At that time, and even more in the succeeding decades of National Party rule, South Africa was run by Calvinist Afrikaners, who banned the teaching of evolution. Even as late as 1994, it was still outlawed in government schools. Smith's experience was a repetition of the reaction that had greeted another famous South African scientist, Raymond Dart, thirteen years earlier, when he published The Taung Child, about his discovery of relics of the first ape man, a true missing link to our past. He too was sent hate mail, warning that he would "roast in the quenchless fires of hell." Letters were written to the papers denouncing the "satanic mischief."

"It is a wonderful piece of irony that so many important discoveries of an evolutionary nature have occurred in South Africa," says Professor Philip Tobias, a former pupil of Dart who today wears his mantle as South Africa's premier paleoanthropologist. "It was also an example of the government's schizoid mind-set. While there was a nationalistic pride that it was within South Africa that the first ape man was discovered, it was only very recently that evolution was permitted to be included—and then only as a voluntary supplement—on the school syllabus."

J.L.B. Smith's studies progressed well. He was thrilled each time he discovered structures that matched fossils millions of years old. The board of the East London Museum, however, was

---

*Margaret Smith apparently deemed this correspondence "irrelevant," and from her deathbed, ordered that the crackpot file be destroyed.

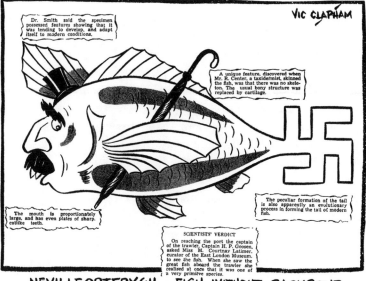

NEVILLEOPTERYGII – FISH WITHOUT BACKBONE

Satirical cartoon of British Prime Minister Sir Neville
Chamberlain as a coelacanth.

becoming increasingly frustrated by their prize exhibit's continued
absence. They sent a telegram to Smith, requesting its immediate
return. The people of East London had been clamoring to see the
fish again, and interested observers had traveled from far afield in
the hope of being able to view it. Although Smith felt he had not
finished his examination, he agreed to give it back. On May 3,
1939, the coelacanth—again with its police escort—returned to its
museum home. The museum was inundated with visitors for the
next few weeks. Smith admitted to feeling a "sense of relief" at its
departure, and Marjorie Courtenay-Latimer was delighted to be
reunited with her fish.

Smith's detailed monograph, comprising 106 pages of text
and 44 photographic plates, was dispatched to the *Transactions of
the Royal Society of South Africa* at the end of June. It was remarkable not only for its detail, but for the absence of a bibliography.

Throughout the entire treatise, Smith did not once mention anyone else's work, something almost unheard of in scientific papers. He and Margaret had five days to recover from six frenzied months of coelacanth activity before their first son together, William, was born. "William arrived two weeks early," Margaret said, "and has been in a hurry ever since."

"My gran," recalls William, "was convinced that I would have no clothes when I was born, while there were those who feared I might be born with scales. Fortunately it turned out fine. Granny saw to the clothes and there were no scales."

Some months later, having fought off an attempt by the museum's board to send her coelacanth to the British Museum, Marjorie Courtenay-Latimer accompanied the fish to Cape Town in a special railway van donated by South African Railways. "That was a great experience," she recalls, "because every time they changed guards, they'd come and say, 'The specimen's quite happy. Resting quietly.' When we got to Cape Town, all the flags were flying. I thought it was to greet the coelacanth, so I waved as we drove through the streets. I later found out that there was some important foreign visitor in town." At the South African Museum, Mr. Drury, their best taxidermist, set about remounting the coelacanth. As no one had seen a live coelacanth, and so had no idea of how it looked when swimming, he mounted it—as Robert Center had done—with the pelvic and pectoral fins pointing downwards, in a distinctly leg-like fashion.

Marjorie Courtenay-Latimer returned to East London on September 3, 1939, the day England and Germany declared war. "Amid the chaos and the fright, all I could think of was that the coelacanth was safely down in Cape Town."

III

INTER PISCES

J.L.B. Smith tried not to dwell on the loss of *Latimeria*'s soft parts, but their absence nagged away at him day after day, until he was forced to acknowledge his growing obsession with finding another coelacanth. "There must be others somewhere," he wrote, "and at the back of my mind 'a cloud no bigger than a man's hand' had formed, the forerunner of the project that came to overshadow all else in my life—the hunt for the home of the Coelacanth."

He started to raise funds for the hire of a suitable coelacanth-hunting vessel to take him to the coral reefs and palm islands of East Africa, where, he was convinced, he would discover the coelacanth's home. He was sure that *Latimeria* was not native to the intensively fished southern African coast—if it had been, it surely would have appeared before—but was a stray that had drifted down the East African coast, carried along by the warm, south-flowing Mozambique current.

The outbreak of World War II, however, put his plans on hold. The world was in turmoil, and until it settled down, he knew he would never be able to find a ship to take him on research trips to the Indian Ocean. He returned to his chemistry teaching, but the illustrated monograph of the coelacanth remained on his desk, a daily reminder of the great task ahead of him. It was not in his nature to give up the search. Where other men might have been content with the fame and respect the first coelacanth had brought them, for Smith it was only the beginning. He *would* find another coelacanth, but for now he had to wait.

James Leonard Brierley Smith was not an ordinary man. He was a brilliant obsessive, devoted to his work, and incapable of yielding either to his own weaknesses—mainly physical—or to the defects of others. His work was his life, and he involved those around him in it to the point where it became their life too.

J.L.B. Smith was born on September 26, 1897, in Graaff Reinet, a pretty town hundreds of miles from the nearest coastline, in the middle of South Africa's Karoo Desert, where his father was the postmaster. Of English, seafaring stock, Joseph Smith introduced his young son to the joys of fishing. "I vividly remember catching my first 'Dassie,' a Bream-like fish, at Knysna," J.L.B. recalled. "This wonderful shining thing I had pulled up from the unknown world below the water had a terrific effect on me, probably more than anything ever since. From then on angling has been a passion, a madness, sometimes even a reproach."

J.L.B.'s mother, Emily Ann Beck, was a beautiful but cruel woman. She was convinced she had married beneath her, and took out her bitterness and frustration on her husband and family. According to *The Dictionary of South African Biography*, "The parents had little in common with their elder son as they were quite unable to understand his sensitivity, his inquiring mind, and his craving for knowledge, education and culture." J.L.B. left home as soon as he could, and soon broke off all relations with his mother and his sister Gladys (to the degree that his son William only recently discovered that he had had an aunt). J.L.B. never talked of his family.

He excelled at the local schools, and in 1912 won a scholarship to the leafy calm of the Diocesan College (known as "Bishops"), Rondebosch, Cape Town, one of South Africa's great private

J.L.B. Smith on a camping trip to Fish River in the late 1920s.

schools, where he was acknowledged to be a brilliant student. One day, he was shooting down a hill on his bicycle when suddenly—too late—he saw a gate across the road that had not been there the previous day. He slammed into the gate and crushed his kidneys. He bled for a year. His lifelong battle for his health had begun.

The First World War broke out during J.L.B.'s matriculation year. He was keen to join up immediately, but as he was still too young, he went instead to study chemistry at Victoria College, Stellenbosch—and came out top of the whole country in his end-of-year exams, gathering a clutch of bursaries and scholarships. He was seventeen, but looked even younger, which indeed he did his whole life, something he regarded as an affliction rather than an asset.

He had intended to go to England to join the Royal Flying Corps as soon as he came of age at the end of 1915, but switched track when the South African prime minister, General Jan Smuts, appealed for his people to enlist with the Allies to fight in German East Africa (now Tanzania). "So instead of learning to soar through the skies, I became an earth-bound, foot-slogging infantryman," he wrote. It was a horrendous period. The men lived and fought an ill-judged campaign, under dreadful conditions. Smith was always a slight man, and despite his considerable sporting prowess (he was the best golfer at Stellenbosch, though playing with only one club, and he later played rugby for his college), he was not strong. In German East Africa, he succumbed to all sorts of grim and debilitating diseases: malaria, dysentery, and acute rheumatic fever. He nearly died in a Kenyan hospital and ended up being shipped back to South Africa with a permanent disability discharge. For the rest of his life, his constant battle with his wartime illnesses would preoccupy him.

J.L.B. returned to Victoria College in 1916 to complete his BA. After another two years, during which time—never one to belittle his afflictions—he writes of being wracked by fever and more often ill than not, he gained his MSc with distinction in chemistry. He worked hard and inevitably came top in exams, but he was not a dull boy. He loved Shaw and Shakespeare, and was blessed, according to his close friend, E. G. Malherbe, with "a fertile imagination."

Malherbe was a fellow member of the "Heavenly Quartet," a group of four smart men, all of whom would go on to great things (Malherbe to the directorship of the National Bureau for Educational and Social Research). In Malherbe's memoirs, *Never a Dull*

Moment, he describes the "most outrageous" student pranks that they devised and carried out. One involved putting all the university clocks forward half an hour; when none of their teachers turned up on time, the students were automatically granted a half-day holiday. According to William Smith, it was the "precise planning" aspect of the pranks that appealed especially to his father.

J.L.B. was awarded a scholarship to study abroad, and in 1919 he went up to Selwyn College, Cambridge, for his PhD. There he carried out research on mustard gases and photosensitizing dyestuffs; he also traveled widely and learned fluent German. J.L.B. returned to South Africa four years later, and was appointed senior lecturer—and later associate professor—of organic chemistry at Rhodes University, Grahamstown. "Doc," as he was known there, soon gained a reputation as a brilliant—if irascible—teacher, completely dedicated to his work. His students still remember his quick movements, the measured, almost pedantic way he delivered his lectures, and his habit of not looking directly at whoever was talking, then swinging round suddenly to hold them in a penetrating stare.

He never, however, let go of his fondness for fishing. It led naturally to an interest in fish, and for him, there was little between an interest and an obsession. He began to spend every available minute trying to identify the fish he found. In those days, there was little in the way of available textbooks, so, characteristically, he devised his own numerical system for identifying and classifying fishes. "It took all my spare time for more than a year, and its compilation involved the writing of more than a million figures, but it worked," he explained. He made contact with the Albany Museum in Grahamstown, and started to publish short (at first) articles on fish in the museum annals. Before long, J.L.B. Smith was a name to be respected in ichthyological circles, and he was sought out by people in need of help in identification.

It was as if he had two full-time jobs: chemistry in term time and ichthyology in the holidays, and to both he devoted boundless energy. He started to explore farther along the south coast, and became the honorary curator of fishes for the smaller provincial museums—including, of course, the East London Museum, with its enthusiastic lady curator—visiting them regularly to classify and investigate strange fishes. He made contact with trawler companies, and spent weeks at a time at sea with them, "often so seasick as barely able to crawl along the slippery heaving decks to scratch among the slimy rubbish shoved aside." He was happiest out in the field, dressed in baggy khaki shorts and sandals, his piercing blue eyes screwed up against the sun, his hair short, in a boyish, almost military cut.

At the same time, Smith started to study fossil fishes. He found this "perhaps the most absorbing of all scientific fields; but my life was already so desperately full that I dared not indulge that desire very far. Nevertheless, those weird creatures of bygone days were constantly flitting in and out of my consciousness, constantly filling me with almost an agony that they had gone forever and could never be seen again." His first forty years were, though he didn't know it at the time, a perfect preparation for the discovery of the coelacanth.

In 1934, a new student entered J.L.B.'s first-year chemistry class. Mary Margaret Macdonald was a composed and popular girl, and a good student, especially of chemistry. She was the youngest daughter of a New Zealand–born doctor, William Chisholm Macdonald, and Helen Evelyn Zondagh, the first woman mayor in the Cape Colony, and a descendant of the Voortrekker leader Johannes Jacob Uys. Her grandmother, at the age of fourteen, had split open a man's skull with an axe at the Battle of Blood River when he tried to crawl under her wagon. Mary was head girl and head scholar of Indwe High School in the Cape Province, chairman of the Debating Society, captain of the netball and tennis teams. She was a talented and accomplished singer and musician, winner of numerous prizes at local eisteddfods, competitive festivals of arts and music. She was determined to pursue a successful career, and her first step was a physics and chemistry degree from Rhodes. There she came into contact with the charismatic Doc Smith.

Initially, she was terrified of him, she later admitted, and it was only in her third year that she realized that he was, after all, human. As she told the writer and photographer, Peter Barnett: "The Professor demands a standard of work and behaviour far above the capabilities of the normal man; many of his former students are today successful men. They often come back to visit him, now as friends, and laugh over all they went through at his hands in their student days. His students fell into two distinct groups, those that liked him and those that didn't. Those who liked him were, in his own opinion, those who Worked. Capitals are used for the word 'Work'. He made things difficult for the others and would waste no time on them."

In Mary's second year, J.L.B. announced that he was going to marry her. "Oh no, you're not," she replied. After she graduated, however, he trailed her to Johannesburg and insisted she marry him. ("And Dad always got his way," said William.) At the time of their marriage, J.L.B. told his young wife that while he couldn't promise her happiness, he could promise that she would never be bored. Their birthdays were both on September 26, and until he was fairly grown-up, their son William believed that all parents shared birthdays.

Margaret Smith* was an intelligent, compassionate woman, who made a friend of everyone she met. She was handsome, with a strong face and deep gray eyes. At twenty-one, she was nineteen years younger than her husband, and perhaps to disguise the age gap, he insisted she wear no makeup and pin her dark hair in a severe bun. He also made her give up music, one of her great passions. "Music stirs the emotions," he used to say, "so it has no place in my life." Margaret Smith became an instant stepmother and great friend to his children by a previous marriage, Robert, Cecile, and Shirley, who were only a few years her junior.†

---

*She started using her second name after her marriage. "While Mary Macdonald sounded good, Mary Smith was grim," she insisted.

†There is no mention of the first Mrs. Smith, née Henriette Pienaar, in the extensive Smith archives. By all accounts, it was an ill-suited match. She was the daughter of a Dutch Reformed Church minister from Somerset West, in the Cape Province, and found her scientist husband hard to fathom. When they divorced, Bob and Cecile—known as Pats—stayed on with their father in Grahamstown, while Shirley went to live with her mother.

She devoted her life to J.L.B. and his work, becoming the perfect foil for her difficult husband. She called him Len (his second name was Leonard, but everyone else called him J.L.B. or Doc), and where he was physically weak, she was strong; she matched his short temper with patience, his distance with warmth. "It was a helluva marriage," recounts William Smith. "They had incredible respect for each other. Each had what the other needed, and they were happy together—as far as Dad could be happy. I am not sure that people like him can be happy; perhaps that is what makes them great."

From the moment of their marriage, they became a team, and until his death, they worked, lived, and dreamed together. It cannot always have been easy: J.L.B. ran his life with an intensity of purpose that permitted neither distraction nor dissension. While Margaret was an extremely intelligent and capable woman, she always had to play second fiddle to her demanding, egotistic husband. "A woman," she once said, "can be independent or indispensable ... but not both. I chose to be indispensable." She proved her mettle during the discovery of *Latimeria*, acting as the rock against which J.L.B. could bash his doubts and fears. After that, through the war years, she joined his fish-collecting trips along the South African coast, becoming as much of an ichthy-enthusiast as her husband.

J.L.B. did not allow the frustration that the war had imposed on his coelacanth plans to hinder his other activities. Every minute of every day was dedicated to work; no one was allowed to slack. Jean Pote, J.L.B.'s secretary from 1966 until his death, describes a "very hard taskmaster. He didn't allow tea breaks; if we wanted tea, it had to be brought to us at our desks, so we could keep working." She recalls how he refused to hire anyone who smoked or wore perfume, and insisted that correspondence be dealt with while it was still fresh. "Letters are like fish," he used to say. "If you leave them more than three days, they begin to smell."

According to his son William, J.L.B. "was very impatient and impossible to live with. He was single-minded to the point of being unbelievable, and while this was a formula for success, certainly, it was not the route to happiness. As a child, living with that intellect was very difficult; I could never win. We were always fighting. If we had not started fighting within half an hour of my getting home from school, Mom would take my temperature. She was the pillow, the shock absorber between us. His behavior, with hindsight, was in danger of breaking me. But it didn't, and I wouldn't change him for anything."

There are numerous examples of J.L.B.'s extraordinary mental powers. He had a photographic memory, and could read sixteen languages and speak eight. When he went to Mozambique for the first time, he learned Portuguese in three and a half weeks, and then proceeded to give an hour-and-a-half lecture without notes. During the war, when he was not fulfilling his teaching responsibilities or hunting for new fish, he managed to produce three chemistry textbooks, which went into numerous editions and were translated into several foreign languages. His intellectual power was also amazingly versatile. "We would go out for a walk," recounts William, "and come downwind of a buck grazing. Dad would motion for us to stop, then he would concentrate on the buck. Immediately, it would start to move around, clearly scared. Then he would turn his mind off again, and it would go back to grazing. It was incredible." Another time J.L.B. recognized, from a distance of fifty yards, a man he had never met, the son of a fellow classmate he had not seen for fifty years. The shape of his skull, apparently, had been a dead giveaway.

But while his mind was strong, his body was weak. At the time of his marriage to Margaret Smith in 1938, he was given less than five years to live. But he was determined not to give in to death, and so he didn't. Instead, he developed a proactive method of fighting his illnesses. He walked for long distances every day (it is estimated that in the last twenty-five years of his life alone, J.L.B. walked the equivalent of twice around the world). He also changed his diet. Using his knowledge of chemistry, he analyzed how the stomach worked, what got digested and where, and came up with one of the first food-combining diets. He refused to mix his proteins and carbohydrates: he never ate meat with vegetables, or bread with butter or cheese. People thought he was crazy. His sandwiches, remembers Jean Pote, used to consist of two pieces of cheese with some apple wedged between them. As Peter Barnett, who accompanied the Smiths on one of their fish-collecting expeditions up the East African coast in the 1950s, explained in his book, "Food is very important to the Smiths, and as they train like prize fighters, there is no room for compromise as far as their diet is concerned. Basically it is an excellent diet, but it eliminates eating for pleasure and places everything on a scientific basis."

By the end of the war, Smith realized he could not maintain his exhausting double life. He and Margaret had become "just machines," working all hours of the day. While he loved chemistry and was a brilliant teacher, his heart was with his fish. In September 1945, he was approached by a complete stranger, Bransby Key, to write a popular book about fishes, and offered an advance payment of £1,000. He immediately handed in his notice to the chemistry department, and soon after, successfully applied to the newly established Council for Scientific and Industrial Research (CSIR) for a research professorship in ichthyology. The university moved him into one of the old military buildings, a tiny wood and corrugated iron shack, which was redesignated the new Department of Ichthyology. From then on, Smith was paid to pursue his passion.

Eight years had passed since the discovery of *Latimeria*, and he was impatient to start his quest for the coelacanth's home, and to find for himself another specimen—with its inner parts intact. He started trying to get together funds for a major pan-African expedition, nicknamed ACME (African Coelacanth Marine Expedition). The committee, composed of South African scientists with a broad range of interests, held endless meetings, and discussions raged as to the proposed breadth of the expedition's aims. But by the beginning of 1948, it was clear that ACME had "fizzled out."

It was not, however, in J.L.B. Smith's nature to give up. He was completely dedicated to his mission, and one way or another, he was going to accomplish it. He tried another tack; if he couldn't go to the coelacanth, he would get the coelacanth to come to him. He had already been contacted by a number of people who claimed to have seen a coelacanth in different places along the southern African coast. In a few cases, the descriptions they sent were likely enough to fill J.L.B. with optimism. Surely, he believed, if he could just spread the word more widely, it would only be a matter of time before another specimen turned up. He persuaded the CSIR and Rhodes to guarantee a reward of £100 to the finders of the first two coelacanths. Marjorie Courtenay-Latimer arranged a special coelacanth exhibition at the East London Museum, and gave the money raised to J.L.B., who used it print thousands of leaflets. These leaflets, with a picture of the fish and the offer of a reward, were translated into French and Portuguese, and distributed up and down the East African coast by every possible means. Smith had high hopes for their success.

£100 REWARD

LOOK CAREFULLY AT THIS FISH. IT MAY BRING YOU GOOD FORTUNE. NOTE THE PECULIAR DOUBLE TAIL, AND THE FINS. THE ONLY ONE EVER SAVED FOR SCIENCE WAS 5 FT (160 CM.) LONG. OTHERS HAVE BEEN SEEN. IF YOU HAVE THE GOOD FORTUNE TO CATCH OR FIND ONE DO NOT CUT IT OR CLEAN IT IN ANY WAY BUT GET IT WHOLE AT ONCE TO A COLD STORAGE OR SOME RESPONSIBLE OFFICIAL WHO CAN CARE FOR IT, AND ASK HIM TO NOTIFY PROFESSOR J. L. B. SMITH OF RHODES UNIVERSITY GRAHAMSTOWN, UNION OF S.A., IMMEDIATELY BY TELEGRAPH. FOR THE FIRST 2 SPECIMENS £100 (10,000 ESC.) EACH WILL BE PAID, GUARANTEED BY RHODES UNIVERSITY AND BY THE SOUTH AFRICAN COUNCIL FOR SCIENTIFIC AND INDUSTRIAL RESEARCH. IF YOU GET MORE THAN 2, SAVE THEM ALL, AS EVERY ONE IS VALUABLE FOR SCIENTIFIC PURPOSES AND YOU WILL BE WELL PAID.

Throughout the late 1940s and early 1950s, J.L.B. continued his work on South African fishes. He came to realize that in order to put them into better context, he would have to study East African fishes as well, so he and Margaret, accompanied by a team of artists, embarked upon a series of expeditions up the coast, from

PREMIO   £ 100   REWARD
RÉCOMPENSE

Examine este peixe com cuidado. Talvez lhe dê sorte. Repare nos dois rabos que possui e nas suas estranhas barbatanas. O único exemplar que a ciência encontrou tinha, de comprimento, 160 centímetros. Mas já houve quem visse outros. Se tiver a sorte de apanhar ou encontrar algum NÃO O CORTE NEM O LIMPE DE QUALQUER MODO — conduza-o imediatamente, inteiro, a um frigorífico ou peça a pessoa competente que dele se ocupe. Solicite, ao mesmo tempo, a essa pessoa, que avize imediatamente, por meio de telgrama o professor J. L. B. Smith, da Rhodes University, Grahamstown, União Sul-Africana.

Os dois primeiros especimes serão pagos à razão de 10.000$, cada, sendo o pagamento garantido pela Rhodes University e pelo South African Council for Scientific and Industrial Research. Se conseguir obter mais de dois, conserve-os todos, visto terem grande valor, para fins científicos, e as suas canseiras serão bem recompensadas.

COELACANTH

Look carefully at this fish. It may bring you good fortune. Note the peculiar double tail, and the fins. The only one ever saved for science was 5 ft (160 cm.) long. Others have been seen. If you have the good fortune to catch or find one DO NOT CUT OR CLEAN IT ANY WAY but get it whole at once to a cold storage or to some responsible official who can care for it, and ask him to notify Professor J. L. B. Smith of Rhodes University Grahamstown, Union of S. A., immediately by telegraph. For the first 2 specimens £ 100 (10.000 Esc.) each will be paid, guaranteed by Rhodes University and by the South African Council for Scientific and Industrial Research. If you get more than 2, save them all, as every one is valuable for scientific purposes and you will be well paid.

Veuillez remarquer avec attention ce poisson. Il pourra vous apporter bonne chance, peut être. Regardez les deux queues qu'il possède et ses étranges nageoires. Le seul exemplaire que la science a trouvé avait, de longueur, 160 centimètres. Cependant d'autres ont trouvés quelques exemplaires en plus.

Si jamais vous avez la chance d'en trouver un NE LE DÉCOUPEZ PAS NI NE LE NETTOYEZ D'AUCUNE FAÇON, conduisez-le immediatement, tout entier, a un frigorifique ou glacière en demandat a une personne competente de s'en occuper. Simultanement veuillez prier a cette personne de faire part telegraphiquement à Mr. le Professeus J. L. B. Smith, de la Rhodes University, Grahamstown, Union Sud-Africaine.

Le deux premiers exemplaires seront payés à la raison de £ 100 chaque dont le payment est garanti par la Rhodes University et par le South African Council for Scientific and Industrial Research.

Si, jamais il vous est possible d'en obtenir plus de deux, nous vous serions très grés de les conserver vu qu'ils sont d'une très grande valeur pour fins scientifiques, et, neanmoins les fatigues pour obtantion seront bien recompensées.

The Smiths' reward poster.

Mozambique to Kenya. Before long, it became clear that the self-taught Margaret was better able to illustrate the specimens to J.L.B.'s exacting requirements than any of the official artists, so she took over as her husband's chief illustrator and remained so for the rest of their working partnership, producing beautiful, accurate watercolors of thousands of different species of fish.

These expeditions were also an opportunity to hunt for the coelacanth. Everywhere they went, they asked about the strange fish, spreading the word and the reward poster throughout the coastal fishing villages. In Bazaruto, an island off Mozambique, J.L.B. talked to a fisherman who claimed he had caught one several years earlier, and described a fish that fitted the coelacanth's description. This was clearly a positive indication of the coela-canth's whereabouts. Sadly, it was the only one.

At about the same time, however, on the other side of the world, the first of a series of clues surfaced that seemed to hint at a secret existence of the coelacanth, far from the area of the western Indian Ocean, on which J.L.B. Smith had focused his search. One day in 1949, Mr. Isaac Ginsburg of the Department of Fishes at the National Museum in Washington, D.C., received a small package in the mail. A woman from Tampa, Florida, had sent him an unusual fish scale, about the size of a silver dollar and different from any scales he had seen before. The woman explained that she ran a small souvenir shop, stocked with fishy artifacts, many of which she made herself from odd bits of flotsam and jetsam— pretty shells and fish scales. One day in 1949, a local fisherman had walked into her shop and sold her a gallon barrel of strange scales. Out of curiosity, she had decided to mail one of them to the museum for identification.

Ginsburg turned the strange scale over and over. He was puzzled; he had never seen scales like that either, and he knew immediately that the fish to which they belonged was not known to live in the Gulf of Mexico, or in any American waters. He wrote back to the souvenir seller, asking for further details, but she never replied and he was unable to track her down. He became convinced that the scale belonged to an ancient fish, probably a crossopterygian and very possibly a coelacanth. In 1949, only one coelacanth was known to science—*Latimeria chalumnae*—and its scales were fully accounted for.

J.L.B. Smith either never knew of the Tampa scale's existence or gave its possible importance little weight. The story was not reported in the press at the time, and as Ginsburg was unable to trace the souvenir seller, he could not be sure of the scale's provenance.

Sea Fishes of Southern Africa was published in July 1949 and sold out within three weeks. In the acknowledgements, J.L.B. paid fulsome tribute to Margaret: "My wife has been my full-time partner from the beginning and has been artist, adviser, buffer, critic and secretary, and is one of the most skilful collectors of fish, with many devices of her own. She has shared with me many hardships and sustained me when my courage has failed. But for her spirit, energy and unflagging enthusiasm this work could hardly have reached completion in any reasonable time." A new edition was rapidly prepared and was issued the following year, again to a tremendous reception.

In the early 1950s, the Smiths embarked upon another series of fish-hunting expeditions, to research further books on the fish of southern Africa. On one, to Mozambique, they took along the young English photographer Peter Barnett, who later wrote a colorful book about his adventure.

In *Sea Safari with Professor Smith*, he describes a grueling expedition with an extraordinary, uncompromising man: "It was never Professor Smith's habit to praise because he said, 'When a motor runs smoothly there is no need for repair, for comment. But when faulty, it has to be licked back into perfect running order,'" Barnett wrote. ". . . No matter what I did, I would become entangled in the complexities of the professor's will. . . . I began to understand a little more about this man, an extreme egotist, sub-

J.L.B. and Margaret Smith fishing on the
Pinda Reef, Mozambique, in 1951.

lime in the certain knowledge of his own intellect."

He was impressed by J.L.B.'s immaculate punctuality. "To such a man as Smith if one arrived a minute over an appointed time, say 3.15 am., then the logical conclusion must be, 'Why not then arrange the meeting for 3.16 am. or 4.16 am.?' J.L.B. used to think that what separated *Homo sapiens* from the rest was time, and he was not convinced women were *Homo sapiens*, because they were not always on time. He was, and he didn't have to wear a watch to be so."

J.L.B. was also frighteningly well organized. "Professor Smith," Barnett wrote, "like Lewis Carroll, numbers every letter he writes, files everything, and had listed every article, down to the smallest box of pins, in the fifty-odd pieces of baggage we took on the expedition. His list, for example, would read 'Box 48' 3/4 down right side 'forceps.'"

As the journey progressed up the coast, Barnett saw that the Smiths "were quite insane about fish, but I soon came to realize that their whole inner selves were focused on the Coelacanth." He told of how they always carried around sheaves of the reward posters, and frequently saw them displayed on notice boards in ports, where the coelacanth was known as the *Dez conto peixe*, the hundred-pound fish. "Both of them were always talking about it to everyone who would listen, which was everyone, because as he said, £100 talks in all languages."

Their general modus operandi, on arriving at a harbor, was to make straight for the fish market, then visit netting boats and fish traps before bargaining for the fish they had selected. When they went out collecting themselves, Margaret would row the boat while J.L.B. issued directions. When they reached what he thought was a suitable spot, he would toss dynamite into the water, and Margaret would dive in to collect the dead fish. J.L.B. didn't like to get wet. "Neither he nor Mrs Smith, I believe, enjoy their expeditions in the accepted and normal way, such as one might experience when hiking," Barnett wrote. "Enjoyment in their busy lives, comes from the inner satisfaction of accomplishments and hardships that they endure willingly in the pursuit of scientific knowledge."

Barnett soon realized that he was expected to push himself to the same degree. He was sitting in a wicker chair after lunch one day when he was pounced upon by the professor: "'Have you any work to do? . . . Relaxation?' he said in tones of stark disbelief. 'But you relax when you sleep, and you'll relax forever when you're dead. In the meantime we've got work to do.'" Everything was accomplished at top speed: "The Professor and Mrs Smith walked fast, the theory being that not only is walking a means of transport, but one must also derive the benefit of exercise from the action so as not to waste time."

Peter Barnett spent several hard months cataloging the expedition, and by the end, he found himself torn between awe and incredulity at the Smiths' dedication and professionalism. One time, J.L.B. rebuked him for not working hard enough, and said that if he didn't pull his metaphorical socks up, he should leave the expedition. Barnett persevered, but when he left them eventually, "The professor did not wave."

In 1952, fourteen years after J.L.B. first vowed to track down another coelacanth, the Smiths embarked upon another expedition up the east coast. They were both becoming frustrated by their inability to find a coelacanth. They had scoured even the smallest fishing villages, but despite seeing their poster displayed prominently, they had encountered no one who recognized the coelacanth, apart from the one possible sighting in Bazaruto. Yet J.L.B. was certain it was there somewhere, waiting to be found—if only he knew the right place to look.

IV

MALANIA ANJOUANAE

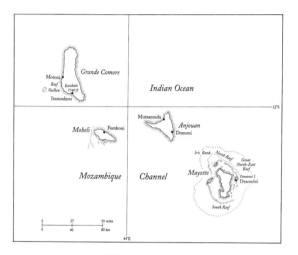

The Comoros.

"I wonder where that blasted fish is. Come on, lass, let's go to the Comoros." The Smiths were standing at Cape Delgado, Mozambique, staring across the Indian Ocean at the point where the prevailing current divides, one section heading south along the coast to Lourenço Marques (now Maputo), Durban, and eventually to East London. J.L.B. had often mentioned his desire to search for the coelacanth in the Comoro Islands, a remote volcanic archipelago at the head of the Mozambique Channel, halfway between Mozambique and Madagascar. The four tiny islands were Arab trading posts for hundreds of years, ruled over by a succession of battling sultans. They became a French colony in 1946, but remained a largely forgotten outpost, unaffected by the twentieth century. J.L.B. was keen to explore their famed coral reefs, but the islands were just too far from the mainland coast to allow easy access by boat, and this time, as on previous occasions, Margaret managed to dissuade him.

The Smiths stopped off in Zanzibar for a couple of weeks of fish collecting on their 1952 expedition. The "perfume isle" was as intoxicating and mysterious as always, but nothing distracted the Smiths from their work. At the end of their stay, they were invited by the authorities to exhibit the more interesting fishes they had found there. The show was crowded with locals, diplomats, and seafaring folk, all keen to see the celebrated professor's catches. Late in the afternoon, a friend of the Smiths arrived, bringing with him Eric Hunt, a keen amateur ichthyologist and captain of a trading schooner. The handsome, charming Hunt, often described as a dead ringer for Errol Flynn (only shorter), was introduced to Margaret, and they started talking animatedly about fish.

As he turned to leave, Hunt picked up one of the coelacanth leaflets. "Do you think this fish might be found in the Comoros?" he asked Margaret. "The Comoros!" she replied. "What makes you ask about the Comoros?" He explained that he often made the journey down the coast, trading local produce, dried fish, and sharks between Zanzibar and the Comoro Islands, a few hundred miles into the Indian Ocean. "I think there is a very good chance that it may be found at the Comoros," Margaret told him. "My husband certainly feels that the Comoros are about the only place left where they may occur." Hunt gestured at the pile of leaflets. "May I take some of these?" he asked. "The Governor of the Comoros would certainly be delighted if one of his subjects could collect a reward of this size."

Two months and many fishing trips later, the Smiths' ship docked at Zanzibar on its journey back to South Africa. Margaret went ashore to visit the bustling market, and as she was returning to the wharf, was hailed by Hunt on his trim schooner, the *N'duwaro* (marlin or sword-fish in Swahili). He recently had returned from the Comoros, he reported, where he had shown the leaflets to the French governor, who had seemed very interested and immediately arranged for them to be distributed throughout the archipelago by native runners. Hunt was full of enthusiasm. It was just the kind of quest that appealed to his adventurous, romantic side, and he was keen to be involved.

Eric Ernest Hunt was an adventurer in the purest sense of the

Eric Hunt and the reward poster, Comoros, 1952. Hunt suffered
from dry skin, and was often seen licking his lips.

word. Born in 1915 to a respectable family from East Sheen, a
leafy southwestern suburb of London, he went to live in East
Africa in 1935. He worked as a motor mechanic for a time, ran a
ferry service on Lake Victoria, and then, at the outbreak of the Second World War, enlisted in the Royal Electrical and Mechanical
Engineers. He served in Abyssinia and East Africa until the end of
the war, receiving citations for his bravery.

In 1946, he embarked upon a new career as a sea trader. He based
himself in Zanzibar, and with progressively larger vessels plied the western Indian Ocean coast and offshore islands with tea and coffee, spices,
cloths, and cloves. He was well liked wherever he went, a handsome,
somewhat shy man with a quiet charm and serious demeanor. His boats

were kept in immaculate condition, and he treated his crew with respect and generosity. Like Smith, his passion for fish arose from a love of angling. Over the years, he became increasingly interested in collecting fish for his private aquarium and studying their behavior. The chance encounter with the Smiths in 1952 and his involvement with the coelacanth provided the kind of excitement he craved.

He plied Mrs. Smith with more questions about the coelacanth: how he would recognize it, and what he should do if he found one and had no formalin. To the latter, she replied that he should salt it in the same way that he salted sharks. As he waved good-bye, he called out, "Okay, Mrs. Smith—and when I get a coelacanth I'll send you a cable." They both laughed.

The *Dunnottar Castle*, the large Union-Castle liner that had carried the Smiths down the coast, docked at Durban on Christmas Eve. It was stiflingly hot, but despite the holiday torpor, a buzz of excitement greeted the arrival of South Africa's most famous scientist. Friends and journalists swarmed aboard the ship to hear news of the expedition. J.L.B. was handed a fistful of telegrams, one of which, he noticed, had been forwarded from Grahamstown and was marked "URGENT" with a red sticker. He was talking to a reporter at the time, and during a lull in the conversation, he casually slit open the urgent message. His reaction was remarkably similar to that on the day he received Marjorie Courtenay-Latimer's first letter. Suddenly he was on his feet, speechless. "Two words stood out: 'Coelacanth' and 'Hunt,'" he wrote in *Old Fourlegs*. Margaret Smith looked up in alarm and took the telegram.

> REPEAT CABLE* JUST RECEIVED HAVE FIVE FOOT SPECIMEN COELACANTH INJECTED FORMALIN HERE KILLED TWENTIETH ADVISE REPLY HUNT DZAOUDZI.

*Hunt had sent an earlier cable to Smith at Grahamstown, which had been intercepted by his secretary, who relayed the information to the professor on the Dunnottar later in the day.

Smith's mind started to fly. He had no idea where Dzaoudzi was, but he knew he would have to get there, and fast, if he was to prevent a repetition of the disastrous loss of *Latimeria*'s soft parts. It was Christmas Eve, four days after the fish had been killed. Time was running out.

Smith asked a young officer to find out where Dzaoudzi was. The officer ran off and was back in a minute with the information that Dzaoudzi was a small islet near the island of Mayotte in the Comoros. It was the Comoros after all! Now Smith had to find a way of getting there; no commercial airline flew to the islands and a boat would take far too long. His only option was to charter a plane.

He stationed himself at the bridge of the *Dunnottar Castle*, commandeered the only phone, and got down to work. First he drafted a reply to Hunt:

IF POSSIBLE GET TO NEAREST REFRIGERATION IN ANY CASE INJECT AS MUCH FORMALIN POSSIBLE CABLE CONFIRMATION THAT SPECIMEN SAFE. SMITH.

He then tried to reach the president of the CSIR, Dr. P. J. du Toit—but found he had already left his office for the holidays and could not be traced. Smith mentally ran through the roll of cabinet members, some of whom he knew, and decided to try Eric Louw, minister of economic affairs: "From the lines on his face his duodenum probably twisted the same way as mine, and I had a suspicion he might look on Christmas time with the same jaundiced eye as myself," Smith wrote. However, Louw was in the United States. Smith tried Eben Donges, an old Stellenbosch contemporary and minister of internal affairs. The post office eventually managed to contact him as he got off a train in Cape Town, but he said that much as he would have liked to help, regretfully he didn't think he would be able to do much from Cape Town on Christmas Eve. He suggested J.L.B. contact the prime minister, D. F. Malan.

Smith wasn't keen to take that route. Several years earlier, he had tried to talk to the then premier, General Jan Smuts, to secure his help in getting to Walvis Bay on the South West African coast. Millions of dead fish had just been washed up on the beach, the victims of a red tide of planktonic organisms, and J.L.B. was desperate to collect specimens. He went in person to Smuts's official Cape Town residence to request an urgent meeting, but Smuts refused even to see him. From that time onwards, J.L.B. had nursed a healthy distrust of prime ministers.

So, instead of calling Malan, Smith tried to contact the ministers of transport and defense and the head of the armed services. He was frustrated at every attempt; they either could not be reached or could not help. By this time, it was Christmas Day, and virtually impossible to get hold of anyone. ("Why on earth did Coelacanths want to turn up just before Christmas?" he wrote.) As in 1939, Smith was unable to sleep; he was tortured by twin fears—that the coelacanth might be decomposing as the days drifted on, or worse still, that it might in the end turn out not to be a coelacanth after all. He was staking his career, he realized, on the word of a layman who had never seen a coelacanth before.

On Boxing Day another cable arrived from Hunt:

> CHARTER PLANE IMMEDIATELY AUTHORITIES TRYING TO CLAIM SPECIMEN BUT WILLING TO LET YOU HAVE IT IF IN PERSON STOP PAID FISHERMAN REWARD TO STRENGTHEN POSITION STOP INSPECTED [PRESUMABLY A MISPRINT FOR "INJECTED"] FIVE KILO FORMALIN NO REFRIGERATOR STOP SPECIMEN DIFFERENT YOURS NO FRONT DORSAL OR TAIL REMNANT BUT DEFINITE IDENTIFICATION HUNT.

J.L.B. was becoming increasingly agitated. "I realise now that I was in a state of mind that is termed 'Possessed,'" he wrote. He feared for his fish—and although he was convinced of his claim to ownership, he knew he would be able to do little if the French decided it was rightfully theirs. A plane was needed—and immediately. Smith was desperate to see the fish for himself.

In his desperation, he knew that there was only one chance left. He had exhausted all other possibilities and would now have to overcome his aversion to prime ministers and beg the help of Dr. Daniel François Malan, the fundamentalist, anti-British, deeply religious prime minister. After J.L.B.'s experience with Smuts, a professed lover of science who would not even see his country's most eminent scientist, let alone help with a domestic flight, he did not have great hopes for Malan, who would have to give so much more—an international plane trip with all the political complications that might ensue. Still, Smith realized he had no option but to try. He engaged the services of the local MP, Dr. Vernon Shearer, and together they telephoned the prime minister's holiday cottage near Cape Town.

Shearer got through to Mrs. Malan, who told him that her husband was already in bed and that she would not disturb him. "10.30pm of the 26th December in the year of our Lord 1952. It was probably the lowest ebb of my life," Smith wrote with typical understatement. "The sands of time were running out, fate was screwing me down to the dregs, wringing out the last drops of my spirit from the rags of my being. . . . What on earth was I to do, for now there seemed no more hope?"

Suddenly the phone rang. Shearer picked it up, said a few words, and called for J.L.B.: "Quick, Professor, Dr. Malan! Dr. Malan wants to speak to you!" J.L.B. took the receiver. "Mrs. Malan here, Professor; the Doctor wants to speak to you." Then the familiar voice came on the phone, speaking in English. "Good evening, Professor, I have heard something of your story, but will you please give me as full a summary as possible." J.L.B. began, insisting on speaking in Afrikaans, even though he stumbled over some of the more technical terms. "I gave him a brief account of the fantastic discovery of the East London fish, the tragedy of the soft parts, my long search, the recent discovery, the heat, the isolation, my fears, and my needs." Smith explained that there was a chance that it might not be a coelacanth, but that there was no question in his mind that it was worth the risk. He said that in his view it was a matter of national prestige—that South Africa had a right and a responsibility to the fish.

He talked for twelve minutes. Malan listened intently, then, when he finished, congratulated him on his Afrikaans. J.L.B. waited. "Your story is remarkable," Malan continued. "And I can see at once that this is a matter of great importance. It is too late to try to do anything tonight, but first thing in the morning I shall try to get through to my Minister of Defence to ask him to locate a suitable aeroplane to take you where you need to go."

"As I put down the receiver," Smith recalled, "I felt dazed, like a man reprieved on the very scaffold, like somebody suddenly jerked from the hollows of hell to a high hill-top in heaven." He was astounded by Malan's positive reaction, particularly to someone who was a scientist with a British name, from such a very British institution as Rhodes.

Later, Smith found out what had happened that night. When the phone rang, Mrs. Malan had judged it better not to wake her husband. However, he had heard the ringing from his bed and called out to ask who it had been. She explained briefly and Malan nodded and said, "This man Smith is well known. Bring me that fish book." Some months before, Smith had sent his *Sea Fishes* book to the prime minister, and by some divine stroke of fate, Mrs. Malan had brought it to their beach house for the holidays. That night, Malan paged through the book and read the coelacanth section. Then he shut it, and tapping the book, said, "The man that wrote this book would not ask my help at a time like this unless it was desperately important. I must speak to him."

Smith worked through the night, preparing lists of things he might need for the trip: food, clothes, a Primus stove, and his collecting kit—a teak box containing trays of tools, spare parts, medical equipment, and fishing tackle. He also managed to find two gallons of formalin, which he added to his traveling gear. By morning he was ready. He went to the dock to wave good-bye to Margaret Smith, who was heading homewards on the *Dunnottar Castle*, then went back to Shearer's house to wait for news from the prime minister. At three-thirty that afternoon, Smith received a message to say that the route had been cleared, and that an air force Dakota would be in Durban at dawn to collect him. He sent a telegram to Hunt:

HOLD ON STOP GOVERNMENT SENDING PLANE.

That night, J.L.B. managed to sleep for three hours, but he was still awake well before first light. He went to the airfield to meet the clearly skeptical Dakota crew. Smith greeted a bemused Commandant Blaauw: "I bet when you joined the South African Air Force you never expected to command a plane sent to fetch a dead fish." He climbed into the unlined hull of the military aircraft with his boxes, and gallons of water, which he insisted were necessary. The plane then took off on the first stage of its journey. It was noisy in the back, and Smith was uncomfortable. He was excited at the prospect of seeing the coelacanth, but also a little nervous: the commandant had informed him that, while the flight had been cleared as far as the Comoros, they had not been able to reach anyone on the islands to alert them of their arrival. Indeed, they had not even been able to find out whether there was a usable landing strip in the Comoros. They would just have to fly there and hope for the best.

The plane stopped to refuel in Lourenço Marques (now Maputo), then headed north towards Mozambique Island. Later, Smith found out that it had not been exactly plain sailing getting permission for landing rights in Lourenço Marques. At 2 A.M. the previous morning, an air force ADC in Durban had called a government official in Lourenço Marques on a crackling line. He requested permission from his opposite number for a military plane from South Africa to pass through the territory and refuel at the air base. "Roger," said the official in Lourenço Marques. "And what is the mission of this flight?"

> Durban: "To get a fish."
> LM: "Have I heard you right, a f-i-s-h?"
> Durban: "Yes, a fish."
> LM: "You mean a thing with scales?"
> Durban: "Roger."
> LM: "Do you really think our government is going to believe that? You must think our guys are stupid—can't you think of

a better story for why you want to cross our territory in a military plane?"

And so the bizarre conversation continued. The Lourenço Marques official eventually agreed to request permission from the Portuguese governor general—a friend of Smith's—and was amazed when he got an immediate green light.

The crew spent the night in hot and humid Mozambique Island, off the northern coast of Mozambique—a familiar hunting ground of the Smiths. J.L.B. passed another sleepless night, and by the time the plane took off before dawn the next morning, he was extremely wound up and nervous. He had staked virtually his whole life on this expedition, and would be ridiculed and ruined professionally if it turned out to be a false alarm. The plane flew low across the Mozambique Channel and soon sighted the first of the four Comoro Islands. From above, they looked like emeralds, smothered in lush vegetation and surrounded by an aquamarine fringe of coral—and beyond, only a few hundred yards offshore, the deep blue of the ocean. Each was mountainous, and on the largest, Grande Comore, Smith and the crew caught a glimpse of Karthala, the largest active volcanic crater in the world. There was little sign of civilization—just some small villages clustered to the sides of the mountains or sheltered beneath tall palm trees along the coast, and in the sparkling sea, tiny, speck-like canoes with two outriggers like open-stretched arms. The most distant of the islands was Mayotte, shaped like a sea horse, with the islet of Dzaoudzi at its side. But even as they passed over, they were still unable to raise the Comoros by radio.

The plane had started to descend, circling the island, when Lieutenant Bergh gave the thumbs-up sign; he had seen the airstrip. J.L.B. looked out of the window, and far below saw a tiny vessel bobbing about in the harbor. He felt a surge of excitement when he realized it had to be Hunt's ship, the *N'duwaro*, with his coelacanth aboard.

The plane made a bumpy landing on the rutted strip, and was enveloped in a tropical downpour. "The rain stopped, as abruptly as if turned off by a tap, the mist parted and figures came running across the flattened coral rag," Smith wrote. "The door opened and through a blast of hot air I saw Hunt's face looking up at me. For a moment I could not speak; then with a rush of pent-up emotion the words 'where's the fish?' burst from my lips like an explosion." Hunt reassured Smith that his fish was safe, and led him to the governor's house. Smith was clearly desperate to see if it was indeed a coelacanth, but Hunt insisted that he greet the Frenchman first. For his part, Governor Pierre Coudert, resplendent in his white tropical uniform, was keen to meet the man who had managed to persuade a prime minister to send a plane to collect a fish.

"I have often suffered from the necessity of paying tribute to officialdom," Smith wrote. "But this was probably the hardest I have ever endured. It was agony and torture, and I raged inwardly, my mind a searing flame. Blast those formalities! I had not endured all that I had been through or come so far to exchange polite words with a Governor at that critical moment. I wanted only one thing, and that was to see the fish, to know if I was a fool or a prophet."

The South Africans were introduced and offered food and drinks that were laid out on a long table. But J.L.B. could endure it no longer; he thanked the governor through gritted teeth and asked most respectfully if they could return after they had inspected the fish. He almost ran to the car, and was soon down at the dock and on Hunt's schooner.

Hunt pointed to a large, coffin-shaped box by the mast, which he ordered to be opened for Smith to see. A sea of kapok was covering the fish. "My whole life welled up in a terrible flood of fear and agony, and I could not speak or move," Smith recalled. "They all stood staring at me, but I could not bring myself to touch it; and after standing as stricken, motioned to them to open it.

"God, yes! It was true! I saw first the unmistakable tubercles on the large scales, then the bones of the head, the spiny fins. It was true! Malan would not suffer for his action, thank God for that! It was a Coelacanth all right. I knelt down on the deck so as to get a closer view, and as I caressed that fish I found tears splashing on my hands and realised that I was weeping, and was quite without shame. Fourteen of the best years of my life had gone in this search and it was true; it was really true. It had come at last."

Smith could have stayed stroking his fish for hours, and the crew would have been more than happy to stay longer on the beautiful and unspoiled islands. But Smith was also nervous that he might yet be prevented from removing his coelacanth. That fear spurred him into action. He carefully lifted the fish out of its box and posed it for photographs. He examined it rapidly, noted its length (4 feet 6 inches, or 1.385 meters), the lack of first dorsal fin, and truncated tail. While it was without doubt a coelacanth, Smith surmised it was of a different species than *Latimeria*.

As he was doing this, Hunt recounted the events leading up to the dramatic rescue mission. By the end of November 1952, Smith's leaflets had been distributed widely throughout the islands. They were greeted with great interest, and the Comorans were amazed at the sum being offered for one fish. On the night of December 20, a fisherman named Ahamadi Abdallah* with his assistant, Souha, took his small galawa—the predominant fishing vessel of the Comoros, a narrow, banana-shaped canoe, roughly hewn from the trunk of a single kapok tree—to fish off the southeast coast of Anjouan, near the town of Domoni. He let out his long hand line, and after a few hours he hooked a big fish, at a depth of 160 meters, which he killed by battering its head. Satisfied with his catch, he returned to his village, and left the fish outside his hut, without scaling or gutting it.

*In *Old Fourlegs*, Smith wrote that the fisherman was called Ahmed Hussein Bourou, but when explorer Quentin Keynes visited the Comoros in 1954, he spoke at length to the fisherman concerned, who assured him that his name was Ahamadi Abdallah, the same name mentioned in the official French report. When Keynes took this up with Smith, he met with the retort: "French administrators were not to be regarded as the most reliable sources of information."

Eric Hunt (second from left), J.L.B. Smith (third from left),
and the crew of the Dakota on Dzaoudzi, Comoros.

The next morning, he took it to the beach to clean. However,
just as he was about to start, he was approached by a local teacher,
Affane Mohamed, who was having his hair cut nearby. Mohamed
(who would later become the Comoran minister of culture) recognized the fish as being very similar to the one on the leaflet. The
instructions were clear: "Do not cut it or clean it or scale it, but
take it at once to some responsible person." Mohamed urged
Abdallah to stop what he was doing, and took him to where the
poster was displayed. The fisherman was at first unwilling to
believe that anyone would be prepared to pay such a fortune for
what he regarded as an almost useless fish, but he was persuaded
at least to see if it was true.

It was known by the bush telegraph—or Radio Cocotier as it
is called in the Comoros—that Captain Hunt was at this time
moored at Mutsamudu, on the other side of the island. It was also
known that it was Hunt who had brought the leaflets to the islands
for Governor Coudert to distribute. Legend has it that Abdallah
lugged his precious 82-pound (37-kilogram) cargo by foot in the

sweltering heat across twenty-five miles of mountainous terrain, but according to Keynes's investigations, he managed to get a lift on a public works truck on its way back across Anjouan from Domoni to Mutsamudu.

Whatever way he got there, by the time he reached Hunt the fish had already started to putrefy. Hunt immediately recognized it to be a coelacanth and started to work out a way to save it for Smith. After a quick inquiry affirmed that there was no available formalin on the island, Hunt ordered his crew to cut and salt the coelacanth, as Margaret Smith had suggested. He promised Ahamadi Abdallah that he would take the fish to Dzaoudzi, the seat of the governor and the only town with the means of international communication, and that he would return with the 50,000 Comoran franc (CFA) reward.

Hunt set sail for Dzaoudzi with the fish and the Anjouan soccer team. When he arrived, he contacted the governor and explained the situation. He also managed to track down the resident French doctor, who gave him formalin to inject into the decomposing fish. Coudert immediately cabled the French scientific base in Madagascar for instructions, but—luckily for Smith— the cable got mutilated in transmission and delayed as a result of the holiday period. Had the French scientists received the correct message, they might well have claimed the fish for themselves. Instead, when Coudert heard nothing, he gave his word that Smith could have it if he came in person to collect it. The coelacanth was won for South Africa.

It was clear that Hunt had stuck his neck out for Smith, in persuading the authorities that the coelacanth belonged by right to South Africa. In return, Smith suggested he christen it *Malania hunti*, after the prime minister and Eric Hunt; but Hunt demurred. He said he would prefer it if somehow the French were honored, since it had been, after all, fished off their territory, and his livelihood depended on good relations with them. So Smith fixed instead on *Malania anjouanae*, to commemorate the island where the fish was found.

Furthermore, J.L.B. said he was prepared to offer £100 for

the capture of another specimen, and that if it was to be caught in French waters, it would be offered to the French.

The governor, Monsieur Pierre Coudert, was delighted at the suggestion of naming the coelacanth after one of his islands, and said the Smiths would be very welcome to return to study the fish on the reefs surrounding the island. According to Smith, the governor and his wife offered the visitors every hospitality. "Madame was concerned at my lack of appetite," he wrote. "Right in front of me was a schoolboy's dream, an enormous cake spread with sticky chocolate icing, the mere sight of which made my liver throb."

Rather before they politely should have, Smith and the crew thanked the Couderts, and after he had sent cables to Margaret, Malan, and the CSIR confirming that it was indeed a coelacanth, they left for the airstrip. They had been on the ground for less than three hours, and Smith had met neither the fisherman who had caught the coelacanth nor the schoolteacher who had correctly identified it. As J.L.B. climbed back into the plane for the return journey, he gave Hunt £200—£100 for the reward to the fisherman and the rest to cover his own expenses—and a handful of newspaper cuttings about the event.

At 10 A.M. on Monday, December 29, the Dakota took off for the return journey to South Africa. They had just got above cloud level when Captain Letley passed a note to Commandant Blaauw, who passed it to Smith: "Managed to intercept a message stating that a squadron of French fighter planes left Diego Suarez before we took off from Dzaoudzi with orders to intercept us and to compel us to turn back." Smith turned pale. He questioned the pilots about their ability to dodge the fighter planes. They said it wasn't likely— the Dakota was too slow and too cumbersome. "Well," replied Smith, "I don't know how you chaps feel about this, but I'm not going back. I don't believe they would dare to shoot us down if we refused to turn, but I would be prepared to chance that rather than turn back." The captain burst out laughing, and it was some seconds before Smith realized that it had been a hoax.

He went to the back of the plane, and to keep his mind diverted from his ears, which were aching, he started to make notes about the experience. At one point, he decided they all needed some coffee, and it was pure good fortune that one of the crew surprised him as he huddled in the back of the plane, trying to fire up his traveling Primus stove. It took the intervention of the captain and commandant—who told him that the fish would be blown to bits if he continued—to stop him. He remained convinced, however, that it would have done no harm. Commandant Blaauw informed Smith that the flight was costing the government £40 per hour. J.L.B. did some rapid sums and calculated that, assuming everything went to plan, by the time the fish was back in South Africa, it would have cost at least £1,000 for the flight. That amount, along with the reward and Hunt's expenses, would surely make the fish one of the most expensive of all time, equal to £12 per ounce in today's prices.

The crew refueled in Lourenço Marques, and at 6:45 P.M. took off again on the last leg of their journey. They had been awake since 3 A.M., but were full of excitement and satisfaction at their successful mission. After an initially cool start, they had come to like and respect the crazy professor. Smith handed them all a piece of paper and asked them to record, in words fit to print, their first thoughts on being called away from their holidays for this extraordinary assignment. The commandant was typically dry and understated, but Captain Letley wrote: "The first time that I knew we were going to fetch a fish (DEAD) was when the Orderly Officer told me. My reply, as you requested, can not be written down." They all said they had enjoyed the experience, nevertheless.

They landed at Durban, exhausted, to be greeted by a battery of flashbulbs. J.L.B. Smith was the man of the hour. He cleared the fish through customs and spoke to the commander-in-chief, requesting permission to take the plane on to Cape Town the next day to show the coelacanth to the prime minister. It was granted. Then the South African Broadcasting Corporation reporter said that the country was waiting for a live radio broadcast from Smith. The regular schedule of programs had been interrupted that evening to make time for him. His first reaction was to refuse—he had not slept properly for almost a week, and had had no time to prepare such an important announcement. However, he remembered the notes he had made in the plane and asked for twenty minutes to organize his thoughts.

The broadcast began, and Smith's confidence grew. His delivery was typically measured, but he could not disguise his emotions as he started to relive the experience. When he told of weeping at the sight of the fish, tears started to fall again. When he finished, he was completely spent. The program was later described as one of the most emotionally charged pieces of broadcasting ever to have been aired on South African radio.

After speaking to Margaret, J.L.B. tried to get a few hours of sleep. He knew that the days and weeks ahead would be as busy and intense as the previous six days, and that he would need all his strength and energy to deal with the inevitable maelstrom of interest in his new discovery. The coelacanth lay beside him, snug in its kapok-lined coffin in Smith's bedroom at the Snell Parade Barracks, while a special detail of Zulu guards patrolled outside.

V

STAR FISH

The dramatic rescue of *Malania* was splashed across the front pages of the world's newspapers. On December 27, the *New York Times* ran a story under the headline "Prehistoric Fish Believed Caught." Three days later, the *New York Herald and Tribune* announced, "Air Race to Save Dead Fish Stirs Scientists Here." The *Times of Malta* reported that "Malan Sends Plane to Collect Reputedly Extinct Fish," while the Karachi *Dawn* exclaimed, "Missing Link Found!" The coelacanth and the Comoros had entered the global lexicon.

On December 30, Smith and his *Malania* were up and in the air before dawn. J.L.B. looked immaculate: his best suit was crisply pressed, but as usual, he wore open-toed sandals, without socks. Even a date with the prime minister couldn't make J.L.B. Smith change his hard-and-fast rule of never wearing closed shoes. The Dakota touched down in Grahamstown to pick up Margaret Smith and William, then took off again, flying over Knysna, where J.L.B.'s eldest son Bob was spending the holidays. "We had no idea what was happening," Bob recalls. "We had no radio and no newspapers. Then a plane flew over the house and Dad dropped a message for us out of the window, written on a board and attached to a parachute made from a sheet."

Showing the second coelacanth to its namesake,
South African Prime Minister D. F. Malan.

As they were starting their approach into Cape Town, Captain Letley passed back another note: "Dr Malan thanks you very much for having taken the trouble to come so far, but he does not wish to see the fish and wishes you a safe return to Grahamstown." Smith was crestfallen. He had come to revere the prime minister, and was eager to show him the extraordinary product of his assistance. His first thought was that it must have been the evolution connection that had upset the Calvinist doctor. As noted in *Old Fourlegs*, he shrugged his shoulders and said, "Okay, let's go to Cape Town anyway for lunch." It was only when he saw the grins on the pilots' faces that he realized he'd been had for the second time. The party landed at Cape Town, and while William was left at the air force base to look at the fighter planes, the Smiths and the coelacanth were driven in a military vehicle to see the Malans.

They were welcomed warmly. The coelacanth, still in its coffin—Smith had refused to show it to anyone before Malan had a chance to see it—was placed under a tree, and the lid opened for the prime minister. He looked at it, then turned to Smith and said, with a twinkle, "My, it is ugly. Do you mean to say we once looked like that?" It was an extraordinary comment from a deeply religious man, an avowed believer in the story of Adam and Eve.

After lunch, the fish was shown to a select group of people invited by the Malans. On his way back to the air base, Smith was recognized by the taxi driver: "My God, sir, you aren't the genilman with the fish, are you? Oh what an honour, what an honour for me and my taxi."

The next day, on New Year's Eve, the "Flying Fish Cart" (as it had been dubbed by the Pretoria News) took off on the last leg of its epic journey. It made a short detour to circle over Malan's house, to drop off copies of the morning papers, as he stood on the lawn waving. A few hours later, the Dakota arrived at last at Grahamstown, where a crowd was waiting to greet it. Among the well-wishers were the mayor, in full mayoral regalia, and Marjorie Courtenay-Latimer, bursting with pride. J.L.B. appeared at the door of the plane, followed by his family, and then by the coffin bearing the coelacanth, which was put on display. "There was great excitement," Marjorie recounts. "We were very thrilled about it, especially that it appeared that they had found the home of the coelacanth in the Comoro Islands." Mrs. Smith was given a bouquet, and J.L.B. presented a copy of his book to the four air force officers and waved them goodbye.* The coelacanth was loaded with Miss Courtenay-Latimer into the museum's station wagon and taken to J.L.B. Smith's laboratory at Rhodes.

*   *   *

---

*To commemorate the fortieth anniversary of the coelacanth flight, the Dakota, which was still in active service, was given a spit and a polish, and flown back to Grahamstown, with a full complement of passengers, including three of the original crew members: navigators Duncan Ralston and Willem Bergh, and radio operator "Vanski" van Niekerk. Two years later, the Dakota was officially retired and donated by the South African Air Force to the SAAF Museum in Pretoria.

The safe arrival home of *Malania* was just the beginning for J.L.B. Smith. At last, he had a complete coelacanth to examine, the result of a fourteen-year search. He was inundated with requests for articles, interviews, and viewings of the fish. This time, there was no war to overshadow the coelacanth story, and the world's press were free to turn the spotlight onto an ancient fish and its eccentric savior. Despite J.L.B.'s antisocial tendencies, he reveled in the attention and was loath to deny interviews to even the most humble of newspapers and magazines. Although he had hardly slept for a week, he could not afford the time to recover. There was too much to be done.

He wrote an account for *The Times* of London, which was typed and sent on New Year's Day (Smith had called the mayor at 5:30 A.M. to ask for two typists), and published the day after. In it, he tried to explain the importance of the coelacanth to mankind and to science. "It is a stern warning to scientists not to be too dogmatic," he wrote. The fact that there appeared to be at least two species of a fish previously and confidently pronounced to be dead revealed the shallowness of man's knowledge of what goes on in the oceans that cover most of the earth's surface. "We have in the past assumed that we have mastery not only of the land but of the sea," he wrote. "We have not. Life goes on there just as it did from the beginning. Man's influence is as yet but a passing shadow. This discovery means that we may find other fishlike creatures supposedly extinct still living in the sea." In addition, Smith stressed how the discovery of a living coelacanth had been of immeasurable assistance to paleontologists. Its remarkable similarity to earlier reconstructions of coelacanths from fossil remains had corroborated the accuracy of their attempts to re-create long-dead creatures from faint imprints on ancient rock. For the most part, he steered clear of the evolution issue—particularly where it related directly to the coelacanth. However, he emphasized that very little was known about the internal workings of creatures from 400 million years ago, and that an examination of *Malania*'s soft parts would teach us something about how the internal organs of those ancient creatures worked, enabling us to go a long way towards clearing up the evolutionary picture. "It may indeed prove to be a sort of H. G. Wells's 'Time Machine,' only always in reverse," he wrote.

The mayor of Grahamstown organized an official lunch for local dignitaries, after which the coelacanth was put on display to the public. Professor and Mrs. Rennie, friends of the Smiths from Rhodes, were invited. They remembered the day vividly. "It was a Saturday afternoon and the town was fairly quiet otherwise, but there was just the biggest mass of people I have ever seen waiting to go into the City Hall," Bee Rennie recalled. "People were converging from all directions. . . . From judges to candlestick makers and heavens knows who. We saw the Judge President sort of pushing his way in, next to Helen Campbell the very short hairdresser."

They described walking along a long passage to the hall where the coelacanth lay in state in its coffin. Grahamstown's only traffic cop, Mr. Archer, his eyes streaming from the formalin, directed the human traffic. "It reminded us of King George V's death in 1936 when we had processed past the catafalque where his coffin was lying in state," Professor Rennie recalled.

Margaret Smith also had vivid memories of the time following *Malania*'s return: "It was the most exciting moment in a life full of interest. For days after returning to Grahamstown, we seemed to walk on air until we came down to earth with a bump when we realised how much there was still to do."

Malania was later taken to the East London Museum, where it spent a few days with *Latimeria*, and to Port Elizabeth for similar dignified occasions. It attracted enormous attention wherever it went—around twenty thousand people were estimated to have seen it in a short period of time. The Smiths received sackfuls of letters and telegrams from well-wishers from within the scientific community and without. One American ichthyologist wrote, "Now I can die happy for I have lived to see the great American public excited about fish." They appeared on television across the world, from America to Japan, Alaska and Timor.

"We were, in fact," J.L.B. wrote, "carried along by a kind of tidal wave we could scarcely control, a wave that went right round the world many times, to its uttermost corners, and the backwash

"If this is the best you can do in 50 million years, throw me back."

keeps on coming back to us even after this long time. In this process an obscure and highly technical scientific term became part of the common speech of mankind."*

When Smith started his closer examination of the fish, he was shocked to note that the coelacanth's brain was missing, and much of the viscera had been badly damaged by its rough treatment. "This was indeed a bitter blow. Once again it was not a

---

*A British MP, in attacking an opponent, called him a "coelacanth" on the grounds that from his long silence in the House, he was surprised to find him still alive.

complete fish," he lamented. He was delighted to discover, however, many extraordinary and wonderful features, including gills that looked like jawbones, thus providing a clue that jaws and gill arches have the same origin. The hollow tube down its back he recognized as a notochord, a forerunner to a backbone and made of cartilage. He confirmed that the limb-like fins had their own internal skeleton, but there was no sign of lungs—instead, a swim bladder filled with fat. And in its intestines he found remains, including scales and the eyeballs of a fish that Smith estimated to have been a good 151 pounds in weight, confirming his belief that the coelacanth was a powerful and successful predator.

The Smiths worked flat-out on his seven-page account for *Nature*, which appeared in the issue of January 17, 1953—only three weeks after the heroic rescue. "It is my great privilege to announce the discovery of a second Coelacanth," he wrote, before thumbing a nose at E. I. White of the British Museum and his theory that the coelacanth was a denizen of the deep. He confirmed his chosen scientific name, *Malania anjouanae*, and stated that it was, in his opinion, a new genus and species,* "and it will in any event not be surprising if further species exist; nor is there any reason why they should not occur in other parts of the world as well," he wrote. "From what can be ascertained the Coelacanths can scarcely be regarded as degenerate fish. They are apparently full of vigour."

Smith noted his preliminary morphological findings, but he soon realized that he did not have the depth of expertise to examine minutely each organ. In a letter to *Nature* dated February 8, 1953, he announced that he would welcome help from scientists in conducting the detailed examination of his fish. Offers from specialists came flooding in, but in his heart, Smith did not want to chop up his *Malania*. While he knew that it would be in the best interests of science, he would have much preferred it if another,

---

*In time, J.L.B. came to realize that *Malania*'s missing dorsal fin and strange tail were probably a result of a shark attack, and not an indication that it was a different species than *Latimeria*.

more complete specimen could be found for science, leaving *Malania* more or less intact.

There were others keen to have a coelacanth of their own. Early in 1953, the *Daily Mirror* reported that the London Zoo was offering £1,000 for a live specimen. The curator of fishes, Mr. Herbert Vinall, apparently had prepared a special large aquarium for it. "We've had all sorts of queer fish in here," Vinall said. "I think we could make this old fossil feel quite at home."

The crackpots started writing again. Letters, mainly from religious people scattered across the world, castigated Smith for his work. One man published a denunciatory pamphlet about his views. There were threats of personal violence; Smith was told it would have been better for mankind had he never been born. He took these attacks with his habitual pinch of salt.

D. F. Malan could not escape castigation for his role in the seemingly apolitical act of saving an old fish. The *Manchester Guardian* printed a barbed editorial directed at the architect of apartheid, pointing out the disparity between his professed political and religious views and his role in fetching the coelacanth. Under the title "Daniel and Darwin," the leader read:

> *It is not often that prime ministers resting from the cares of office are asked to interfere with their holidays in the interests of ichthyology. Dr Malan, it is reported, rose to the occasion with the disinterestedness of the true scientist. But did Dr Malan realise what he was doing? The Coelacanth is believed to be a link of evidence which might prove man's evolution from the fish stage. Dr Malan has been a minister of the Dutch Reformed Church, and only last September the Dutch Reformed Churches of South Africa complained of an exhibition at the Transvaal Museum which purported to trace man's evolution from the apes.*
>
> *The theory of apartheid and the superiority of White over Black would take a nasty blow if it could be proved past doubt that all men, Black and White, have a common ancestor—and an even nastier tumble if it could be shown that the common ancestor is the Coelacanth, a mere fish.*

*   *   *

André Lehr presents the £100 reward to *(from left to right)*
Ahamadi Abdallah, Souha, and Affane Mohamed.

While Smith was grappling with the delicate inner parts of *Malania*
and basking in international fame, Eric Hunt, who had saved the second coelacanth for him, was having an altogether stickier time. He was
proud to have been the agent of Smith's success, but he also foresaw
problems about how the French were going to react to what they
would inevitably regard as their humiliating failure to recognize the
treasure on their own doorstep. Hunt wrote to J.L.B. from his boat the
day after Smith's flying visit, describing lunch at the governor's residence, during which Coudert had confided that he had already been
on the receiving end of some "blunt words" from the French Bureau
of Scientific Research in Madagascar about how he had stood by Hunt
and allowed the coelacanth to go to South Africa. He believed that
they had held back from claiming the fish for themselves only because
they were skeptical of Hunt's identification.

Hunt described the ceremony in Mutsamudu in which the
fisherman, Ahamadi Abdallah, was presented with his £100 award.
According to custom, he shared a third part with his assistant (had
the boat not been his own, he would have given a further third to
the owner), but still, it was an enormous sum of money for a poor
fisherman, and what was more important, the public reward con-

ferred upon him a status above that of his peers. Ahamadi Abdallah told Hunt that yes, the fishermen knew of this fish. It was called *gombessa*, and was caught occasionally. *Gombessa* was not good when fresh, but could be eaten salted. Abdallah confirmed that he had caught the *gombessa* at a depth of 20 to 30 meters, about 200 meters from shore.

Furthermore, Hunt reported to Smith, he had heard of another specimen, which had been caught recently near Mitsamiouli, on Grande Comore, the largest of the Comoro Islands, but which had been thrown away when the local Muslim priest designated it unfit to eat. The scales were used, Hunt was told, for roughing up the inner tubes of bicycles when they got punctures.* And there were rumors of more coelacanth catches, Hunt said, that he was inclined to believe.

He again thanked Smith for naming the fish after Anjouan. "The Govt have always been very good to me here that I feel my insisting on Anjouan being included, was in a small way showing my appreciation of them. It is certainly a temptation to have such a rare discovery named after oneself, but in this particular case I felt my allegiance to the islands which have been my sole trading area for 4 years was stronger." He asked Smith to send newspaper cuttings relating to the coelacanth to his father in London.

Two weeks after Smith and his fish had left the Comoros, Hunt was dealt a bitter blow when the *N'duwaro* was hit by a cyclone and sunk. Hunt returned to Dzaoudzi to await the arrival of a "Lloyds chap" to write off his boat. The storm had devastated the island, he reported. Dzaoudzi, "which was in a way quite pretty, now looks as if an Atom bomb had hit it." Houses had crumbled,

---

*Experiments by coelacanth guru Robin Stobbs, however, indicate that the scale is not strong enough for this to be the case. "They might well roughen up a slab of frozen margarine but would be incapable of abrading a bicycle inner tube or of smoothing a piece of wood," Stobbs maintains.

and the streets were strewn with branches, rubbish, and pieces of crumpled corrugated iron.

A French scientist was now installed on Anjouan, with orders to stay until he got another coelacanth, Hunt reported in another letter to J.L.B. Smith. The governor had come in for a considerable amount of criticism for allowing Smith to whisk the fish—caught by a French subject in French territorial waters—from under French noses. Long and furious articles had been published in Paris newspapers, railing against the "theft" of the coelacanth. "Mind you I can quite sympathise with them and see their point of view," Hunt wrote. "I am quite sure you would be furious if a French schooner type such as myself had got one in SA and sent it to France."

Hunt's letter crossed one from Smith, sent to Zanzibar, in which he described the events following his return to South Africa. He wrote of the broadcast, which had been played across the world by the BBC. "I hope you heard it," Smith wrote, "because it paid full tribute to your ingenuity and determination. . . . [a] Mrs Waddington wrote to say she had heard my broadcast in England and she writes, 'We are so glad that it was Eric. If you wanted a Dinosaur he would get you one.'"

He suggested that Hunt might now find it more profitable to act as a coelacanth guide for all the expeditions that were bound to converge on the islands than to continue hard trading. Indeed, Smith wrote, he and Margaret were most keen to revisit the Comoros later in the year, to investigate the "great treasure house of fish-life.

"Science is grateful to you for your part in this important discovery, and I hope you will use your knowledge to promote others," Smith concluded. Margaret Smith sent her own letter, in which she congratulated Hunt: "As long as the human race continues, your name will be linked with one of the most thrilling stories of our age. It is very gratifying to me that my confidence in you was so well founded and that you were in no way to be overawed. . . . You have certainly attained one of your great ideals: to put the Comoro Islands on the map. After this we expect them to receive a considerable amount of attention from the outside world."

In France, the anti-Smith sentiment was heating up, and

unfortunately for Hunt, stranded in the Comoros after losing his ship, he was well placed to take the flack. Pressure was put on Governor Coudert to do something. It was he, after all, who had allowed Smith to whisk the coelacanth away from French territory, and he now was forced to take action to save his career. He turned on the nearest scapegoat and proceeded to confront Hunt and to write to Smith contradicting Hunt's version of events, and presenting a very different scenario.

At this point, Hunt's gratitude towards the French started rapidly to fade. Obviously upset, he wrote to Smith: "I was furious and told the Governor that over the whole affair I had not sought to make a personal gain out of it, and therefore the least I could expect was the credit due to me for having started the necessary propaganda, and personally guaranteed the payment of £100. Also, having been the only person concerned with the preservation of the fish. *Nobody bar myself and my crew who took turns with me in injecting the fish, had anything whatsoever to do with it.*"

Hunt laid out in detail the steps he had taken to secure the coelacanth, and the governor's cooperation in the process. "Paris was furious that the fish had not gone there. You say ultimately it would have come back to you. I am not in a position to judge, I don't know. But I do know this, that Paulion who I talked with in Tananarive was furious with the Governor . . . [who] in order to extricate himself a little wrote his letter to you cutting me out of the affair almost entirely, and more or less making me out to be only the transport for the fish from island to island. Claiming that he had handed the fish over to you in the name of France and that in Anjouan the Govt. had taken charge of the fish and treated it before you came into the picture.

"I am really quite upset over all this, as from being a friendly affair between ourselves, it has turned into quite a political one," Hunt wrote.

He explained that the French had made him sign an official version of events that differed radically from the true story. There were three versions, Hunt explained: the official story sent to France before Smith's arrival (which corroborated Hunt's account);

the doctored version in response to pressure from Paris—which effectively cut Hunt out of the story and which he refused to sign; and the compromise version, which still put a heavy French spin on events. "The Comoros are my means of earning my living," Hunt wrote, "and studying fish, and fishing my hobby. I must by this reason remain on good terms with the French authorities, and to that end I have signed the latest official story of the Coelacanth. . . . I wish now that I had accepted your offer of naming the fish after me, it would certainly have guaranteed me a certain amount of credit, which I have not received.

"Without being too much of an egotist, I think that you appreciate that if I had not met your wife in Zanzibar there would have been no *1952 Coelacanth for anybody.*"

Smith replied in lofty tones, reassuring Hunt that while he was "obsessed with what those Authorities have said, this is not what has been published all over the world. You need not feel you have not been given ample credit. You have. . . . You have actually had more credit for asking that the fish should not be named after you, but after Anjouan. . . . Both my wife and I have seen that your ability was appreciated. . . . Mrs Smith has repeatedly said in public that she always felt that you were the most likely layman to be able to spot an animal of that kind."

He played down the actions of the French, suggesting that they did not "really understand the full position. They were so astounded at the world's reaction to this that they felt they had been deprived of something important," he wrote. "What they did not realise was that if anyone else had got another Coelacanth, it would certainly have been interesting, but the terrific reaction everywhere was not by any means to the Coelacanth alone, but of the remarkable circumstances surrounding it. First of all perhaps most interesting to the world at large was the fact that this was the direct result and culmination of my 14 years' research. Secondly, there was your part in it in direct relation to us and to the long search." Malan's intervention, too, accounted for some of the interest, Smith conceded. "None of these things would have been attached to the discovery by any other person. Furthermore, had the French seized the specimen the world would have known that they had got it not by their own efforts, and they would certainly not have been admired for their action."

Smith apologized that he was not in a position to reimburse Hunt for outstanding expenses—the telegraphs and time lost in waiting for the plane—over and above the £100 he had already given him, but added that he hoped to raise funds for an expedition to the Comoros later in the year. Sadly, that was not to come about. The Smiths and Hunt stayed in contact with each other, but they were never to meet again.

Hunt's involvement in the coelacanth episode—notwithstanding the abuse he had had to take from the French for J.L.B. Smith's dramatic appropriation of what they saw as a French fish— further stimulated his interest in the aquatic world. After the *N'duwaro* was wrecked in 1953, he bought a new, larger, hundred-foot schooner in Naples, which he named *Hiariako* (Swahili for your choice). Two years later, he married Jean Fowler, a Scottish girl fourteen years his junior, in Tananarive, the exotic French colonial capital of Madagascar. She frequently accompanied him on his voyages, but Hunt still found the life of an ocean trader incompatible with that of a husband. Fish had, by this time, evolved from an interest into a passion. He not only kept an aquarium at home and on his ship, but had started to study them in detail, and to become involved in the conservation of the marine environment. He planned to sell his ship and settle in Madagascar with Jean, and to set up a business collecting exotic fishes for aquariums around the world.

On April 9, 1956, the Hunts set off from Zanzibar in search of fish specimens and cargo. Eric had confided to a friend before he left that it would probably be his last cruise before he sold up. They spent a few days in the Comoros, where they dropped off some cargo, then sailed for Majunga, a pretty town on the north-west coast of Madagascar. There the Hunts collected fishes, which they transferred into the onboard aquarium. Jean then left to search for Madagascan lace plants and visit friends in Tananarive, while Eric headed back to the Comoros with another load of cargo. He sent her a telegram telling her to fly over and join him in Dza-oudzi, where he would meet her at nine-thirty on the morning of May 3. He never arrived.

According to an official inquiry, the *Hiariako* set sail from Majunga on schedule in the early hours of May 1. With Eric were fourteen crew and eleven passengers. Before long, they found themselves being jostled by large waves and swept along by strong winds. By the next morning, Hunt had to admit he was lost. All that day and night, they searched in vain for a sighting of land, and at 4 A.M., while Eric was asleep, the vessel struck the Geyser Reef and slowly ground to a listing halt on the treacherous coral, only eighty nautical miles from their destination.

The crew tried desperately to refloat their ship. They jettisoned all the cargo, but water continued to pour into the unfortunate *Hiariako*, and Hunt was forced to give orders to abandon ship. He, a French passenger, and the cook climbed into a small sailing dinghy; the senior members of the crew, with three Comoran women and three children, took the ship's boat; while the remainder spread themselves across a wooden raft and another makeshift raft constructed of oil drums and hatch covers. Each had a little fresh food and water.

At first, Hunt attempted to tow the other vessels to shore with his small outboard engine, but he soon realized that wasn't going to work. He decided to hoist sail and set off with the Frenchman and the cook to try to raise help from Mayotte. He was never seen again. A vicious storm engulfed the sea, and probably took the people on the rafts with it. The ship's boat managed to outweather the storm, but the nine people on board soon ran out of food and water, and within days two of the children and one of the Comoran women had died. Fifteen days later, on May 20, the survivors were spotted by Comoran fishermen in three small outrigger canoes, and taken back to the safety of Grande Comore.

On May 24, Hunt's dinghy was found, on its side and empty, fifteen miles west of Moheli, the smallest of the four Comoran islands. The engine appeared to have been torn away from its bracket, and there were no signs of its human occupants, apart from the floating remains of two Mae West life jackets some three hundred feet away. The bodies of Hunt and his two companions were never recovered.

The shipwreck caused excitement in the British press when an erroneous report in the *Daily Telegraph*, quoting unnamed British Intelligence sources, suggested that Hunt had been killed during a daring attempt to spirit the leader of the Greek community in Cyprus, Archbishop Makarios, away from his Seychelles island exile. Hunt's family were terribly upset by this slur on Eric's character, and numerous angry letters were exchanged with an unyielding *Daily Telegraph*. The official report into the accident, however, makes no mention of kidnapping missions, and concludes that it was an accident, and that neither the captain nor the crew were to blame.

In an obituary titled *An Aquarist Dies at Sea*, Henry Nichols FAI wrote: "Eric Hunt will be missed deeply by all his fisherman friends, far and near; but also his departure is inevitably going to leave a very empty spot in the lives of simple, humble black and brown people along the central East African coast from Pango to Lourenço Marques and on many of the offshore Indian Ocean islands." Nichols wrote of how Hunt helped the locals with money and goods as well as his friendship and trust. "He was an artist," he concluded, "in the best sense of the word. A man who lived the life he loved and loved the life he lived til its last roaring, foam-swept moment—trying to help somebody."

J.L.B. Smith was left bereft when in 1953 the French refused him permission to return to the Comoros. "The global reactions to the extraordinary culmination of my long search for the Coelacanth had profound effects in France, where widespread and somewhat hysterical propaganda in the press aroused public feeling," he wrote. "As a result, there was widespread agitation, and it was urged that the French Government should demand the surrender of the Coelacanth to France." A diplomatic incident was clearly brewing: the loss of *Malania* still rankled, and so it was no surprise when on November 9, 1953, Paris issued an edict: "Only French scientists will be allowed to search for the coelacanth off the French Comoro Islands, in the Indian Ocean between Mozambique and Madagascar, for the rest of this year. French authorities there have declared a complete ban . . . on expeditions by foreign scientists."

The coelacanth had become a French fish.

VI

NOTRE COELACANTHE

The French were determined to make up for lost time, and to claim the coelacanth as their own. They knew they initially had underestimated the excitement that *Malania*'s discovery generated. Both the scientific world and the public at large were ablaze with interest in the strange and ancient fish, and the French wanted to be a part of it. Even Hollywood latched on: *The Creature from the Black Lagoon*, featuring a finned monster that emerged from the sea, apparently was inspired by the discovery of *Malania*.

Dr. Jacques Millot, a spider expert with an uncommon resemblance to a kindly hobbit, at that time headed up the French research facility in Tananarive, Madagascar. He took charge of coordinating the search for a French coelacanth, and for the next two decades, foreign scientists continued to be excluded. The glory of unraveling the intricate inside workings of the world's most celebrated fish was to be French alone.

Millot installed one of his colleagues, Pierre Fourmanoir, in the Comoros with the aim of instigating a concentrated fishing effort among the local fishermen. Needless to say, the £100 reward they were promised for what had become known as *"Le Poisson"* was sufficient spur to galvanize their efforts.

On September 26, 1953, a 51-inch, 87-pound coelacanth was caught off Mutsamudu, Anjouan, the far side of the island from where *Malania* was fished up. The fisherman, Houmadi Hassani, recognized it immediately, and quickly took it back to his house. His wife stood guard while he ran for help to the French doctor, Dr. Georges Garrouste, who had been equipped by Millot with a special Coelacanth Preservation Kit. He collected the fish in his ambulance and injected it with seven gallons of formaldehyde. The next day it was flown by a special aircraft to Millot in Tananarive. The French were overjoyed. Pictures of the specimen adorned the front page of *Le Monde* under the jubilant headline *"Notre Coelacanthe!"*

J.L.B. Smith was on the island of Bazaruto, off the coast of Mozambique, at the time. He was in the process of examining a huge rock cod, which had been found in very shallow water, when he was approached by a Portuguese-speaking Chinese man. This man had heard a report on the radio the previous night about a large fish with a strange name that had been caught by the French, somewhere near Madagascar, he thought. Smith's name had been mentioned, which was why he had approached him. Smith interrogated the man, and established that it was definitely a coelacanth, but beyond that he could elicit no further information. He searched for a recent newspaper, or someone else with a wireless who had heard the broadcast, but it was not until he got back to civilization a week later that the report was confirmed: the French had caught their own coelacanth in the Comoros.

"I shall always remember the sensation of terrific relief this gave me, as if a crushing burden had been lifted from my mind," he wrote in *Old Fourlegs.* "So it was the right spot, they were there! I could see the end of this great strain. I could keep my *Malania*, and it would be only a matter of time now before all those high specialists were each and all wresting the secrets of the life of the long ages past from the tissues and structures of fresh coelacanths." Smith immediately sent a telegram of congratulations to Millot.

The small volcanic islands at the western edge of the Indian Ocean were, in Smith's mind and the minds of most people, confirmed as the home of the coelacanth. The aim of his fourteen-year search had been fulfilled. What nobody considered at the time was that the Comoros are relatively young islands: the oldest, Mayotte, was pushed up from the sea only 5.5 million years ago; Grande Comore, a mere 130,000 years ago.

The year 1954 brought a bumper harvest of coelacanths for the French. Their second specimen—the fourth to have been officially recorded—was taken near the village of Iconi in Grande Comore, indicating that coelacanths lived off at least two of the four islands that make up the Comoran archipelago. Its capture on January 29 was followed by a similar dash to box and transport it. "It was very exciting, rushing to preserve it, to build a case, to order a special plane from Madagascar," noted an American reporter from *Collier's* magazine.

"We finished up tired and proud at 4 P.M., then a man staggered in with an even bigger coelacanth," recalled Maurice Rex, the administrator of Grande Comore. Specimen Five, caught near the small village of Mandzissani, was boxed up and sent with Specimen Four to Millot in Madagascar.

Suddenly there was no shortage of coelacanths to dissect. Up to this point, no photographs had existed of coelacanths that were not damaged and decomposing, or stuffed, rigid, and brown. For the first time, Millot saw what Marjorie Courtenay-Latimer had described as "the most beautiful fish." He and his assistant, Jean Anthony (a shy comparative anatomist), were soon elbow deep in coelacanth viscera. They started work on what would become the three-volume, profusely illustrated *L'Anatomie de Latimeria*, the definitive study of the anatomy of the coelacanth, and probably one of the most detailed ever examinations of a fish, which would take them eighteen years to complete.

Much already had been written about the coelacanth's external features by the time Millot got to work, mainly by J.L.B. Smith in his detailed monograph on *Latimeria*. The world was familiar with its hard, mottled scales, its wide mouth with those generous lips, its large eyes. Close attention had been paid to the skeleton of its limb-like fins, which are unmistakably similar to early tetrapod limbs, and to its unusual fringed tail, seemingly a continuation of the body, with its small "puppy dog" extension. Now, with the benefit of a complete coelacanth, Millot and his team were able to discover the detailed inner workings of this relic of ancient times.

Several features exhibited marked similarities to those of living tetrapods: the coelacanth's inner ear was found to have more features in common with frogs than with other fishes; its gill arches were similar in shape to jawbones, with gill rakers resembling teeth; and its blood was found to contain large red cells, similar to that of amphibians and lungfish. All of these confirmed that the coelacanth was indeed very closely related to the first fish that climbed out of the water to found the greater part of the animal kingdom on land.

But there also emerged information that for a short time appeared to indicate a kinship with sharks, among the most primitive of fishes. The coelacanth's blood, for instance—like that of sharks—was found to be rich in urea, a nitrogen-based chemical that is the main constituent of mammal's urine. Instead of a backbone or ribs, the coelacanth has a fibrous notochord—hollow spinal column—similar in composition to the cartilaginous disks of sharks. The coelacanth's notochord is filled with a unique oily fluid. Its V-shaped, symmetrical heart is also like both a shark's and that of most vertebrate embryos; and Millot found a spiral valve in the coelacanth's large intestine similar to that found in sharks, which allows them to digest their food very slowly and completely, and to go for longer between meals. This is clearly an adaptation to living in food-poor areas.

Indeed, that the coelacanth appears to be singularly well suited to its environment is not really surprising in such a successful survivor. It has a very slow metabolism, which means it can conserve energy when food is scarce. It also has a small gill area—less than that of any fish of comparable size, and one hundredth of that of a tuna—so that it absorbs oxygen at a slow rate, making it well adapted to living in deeper, cooler areas, where the water has a higher concentration of oxygen, and where there are fewer predators and less competition for available prey. The coelacanth's slender, tubular swim bladder is filled with fat and other noncompressible body fluids (unlike the hollow, gas-filled organs of bony fishes), which gives it neutral buoyancy, allowing it to hover comfortably at great depths without having to move and thus expend energy. It has a reflecting layer behind the retina of its eyes which, as in cats' eyes, gives it good vision in dim light (and conversely, as the eye contains no melanin, blinds it in bright light). The Comoran fishermen describe coelacanth eyes shining like "lights or burning coals."

The coelacanth has a pair of features thought to be unique among living creatures. Its brain—which is primordially simple and very small, weighing about 1/15,000 of its adult body weight and occupying less than 1/150 of the brain cavity, about the same size as a grape—lies behind a hinge-like intracranial joint that separates the nasal organs and eye from the ear and the brain. This joint is considered to be very primitive, and is thought to allow the fish to open its jaws wide and thus increase the power of its bite. Similarly unique are the six jelly-filled cavities embedded in the snout. Scientists have christened this feature the "rostral organ," and speculate that it is a highly complex electro-detection system, used to locate prey.

To the scientists then studying the coelacanth, its inner workings were like a series of signposts pointing in different directions. Part of its appeal was undoubtedly its well-publicized propinquity to the main trunk of the evolutionary tree. Millot and his team clearly had hoped that the availability of a modern descendent to study would help to clear up the ongoing and often fierce debate as to which of the lobe-finned Sarcopterygii was the first to crawl out of the water and colonize the land. Their early results, however, led them around in circles, and after a few years, the missing link between fish and land-dwelling creatures had still to be found.

*   *   *

Among the earliest contenders for the position of possible link
between aquatic and terrestrial creatures was the lungfish. The first
living specimen was discovered in the Amazon by a Viennese naturalist, Johann Natterer. He returned to Europe in 1836 from eighteen years of travel in Brazil with an enormous collection of exotic
creatures that increased the inventory of the Imperial Museum in
Vienna sixfold. His prize discovery was a fish-like beast, about two
feet long and eel shaped, with both gills and fully functioning
lungs. The locals called it *caramuru,* and Natterer coined the scientific name *Lepidosiren.* He wrote in a monograph published shortly
after his return that it was a "new species of animal of the family of
fish-like reptiles (Ichthyodea). It deviates in all its characteristic
details so significantly from other representatives of that group [of
reptiles], and resembles a fish so closely in its overall formation
that even the most experienced natural investigator can be mislead
[sic]." Time would prove that it was Natterer himself who was
thus fooled.

One year later, a similar lungfish was brought back from the
Gambia River in West Africa by Englishman Thomas Weir, along
with a second, dried specimen enclosed in sun-baked clay. This
lungfish was known locally as *comtok* and christened scientifically
as *Protopterus.* It was soon discovered that, during the dry season,
both species curl up and go into estivation (summer domancy) in
round mud cakes, and only reawaken when the rains come and
soften their muddy nests.

The living lungfish were discovered at around the same time
as the first coelacanth fossils. Scientists quickly concluded that the
two species were closely related—and both were regarded, in their
time, as the cause célèbre of icthyology and evolutionary theory.
These unusual fish, too, were greeted in the beginning with a
degree of confusion. Natterer's initial monograph on the Amazo-
nian lungfish had sparked a fierce debate among scientists as to
what exactly they were: fish or amphibian? The question, according to one scientist, had "singularly embarrassed taxonomists."
Under most classifications, fish had scales and gills, while amphib-

ians—with naked skin—possessed the ability to breathe. What did that then make *Protopterus* and *Lepidosiren*, with their scales, gills, *and* lungs?

Contrary to Natterer's initial diagnosis, majority opinion settled on the side of them being breathing fish (rather than scaly amphibians), possibly somehow related to the first tetrapods. The problem was that they clearly were not walking fish, as neither had fins to speak of—just thin, string-like appendages—and they swam in an eel-like manner, their whole body squirming. Scientists, led by Darwin and Wallace, were convinced that the true stem of land-dwelling creatures possessed both the ability to breathe air and the legs to walk on land. The lungfish fulfilled their first criterion, but appeared to fail miserably on the second. Indeed, up until 1839, when the first coelacanth fossils were unearthed, there still hadn't appeared any fossil evidence of a fish with limb-like fins. To fill this "missing link," the anatomist Karl Gegenbaur had invented such a fin with "leg possibilities"—his students christened it *Archipterygium gegenbauri* (Gegenbaur's archaic flipper)—evidence of which, he prophesied, would some day be found in a fish-like creature that could breathe air.

The coelacanth fossils clearly showed the lobed fins, but there was no visible evidence of anything that could be construed as a lung; and until a living representative was found, there was no way of knowing whether these fossil fish had had the ability to breathe. Then in 1869, an event pushed the lungfish once more into the spotlight. An Australian bushwhacker named William Forster moved to Sydney from his cousin's farm near the Burnett River in Queensland, and soon after arriving in the city, he visited the Sydney Museum. He fell into conversation with the curator, Gerard Krefft, about the unusual animals found in Australia. Forster asked why the museum had no example of a strange fish that lived in the Burnett River. This fish, Forster explained, was known locally as the Burnett salmon. Krefft was interested and asked him to describe the fish. It was about five feet long, Forster said, round like a fat eel, with large greenish scales and four strong fins. Krefft said he had never heard of a fish that fit that description, but would very much like to see one. Forster promised to write to his cousin, asking him to send a Burnett salmon to the museum.

A few weeks later, a large barrel arrived for Krefft containing several specimens, strongly salted to prevent them from decaying. He took out a five-foot-long fish to examine. Forster's description had been pretty accurate; Krefft saw the scales, and the four unusual fins that looked more like paddles. He was most struck by the tail, which was unlike any fish tail he knew: it looked more like an extension of the body, fringed with fin material. But it was when he carefully prized open the fish's mouth to look at its teeth that he was stunned. He immediately recognized the four large teeth, which were joined together like a rooster's comb, to be remarkably similar to fossil remains of teeth found in ancient rock deposits. No living fish he knew of had teeth like that. Nobody had worked out what the ancient creature belonging to those teeth had looked like; it had even been suggested, by the paleontologist James Parkinson, that they represented the "digitated termination of the sternal plate" of a tortoise. Louis Agassiz—the same Agassiz who had named the first coelacanth fossils—had christened the owner of these teeth *Ceratodus* (the horn-toothed).

Krefft delved into the archives to search for further clues and was astounded at what he found. Nowhere was there mention of a living fish with fins matching the description of the Burnett salmon's; the only similar sort of fins he could find were Gegenbaur's imaginary archaic flippers. Could this fish lying in front of him on the dissecting table possibly be the missing link, and not a mere fossil, but the real thing? Krefft was filled with excitement. He continued his dissection and was overjoyed to discover that the so-called salmon had, as well as gills, a single lung. He came to believe that he had stumbled upon the original primitive lungfish, a forefather of *Protopterus* and *Lepidosiren,* and quite possibly of all mankind. He christened it *Ceratodus forsteri.* When a complete fossil later was found that matched exactly Agassiz's *Ceratodus,* and which was subtly different from the modern Burnett salmon, the name was changed to *Neoceratodus* (new *Ceratodus*).

For a long time, *Neoceratodus* was a zoological sensation. It was to the nineteenth century what *Latimeria* is to the twentieth, and it was as exhaustively dissected, examined, and debated. Ernst Haeckel, who succeeded Darwin in being known to his contemporaries as the greatest living zoologist, took a great interest in the Australian lungfish, and became determined to investigate its ontogenesis.

Ontogeny was, in the late nineteenth century, a fashionable study. It looked at the development of a species from fertilization to birth. Haeckel, an impressive and erudite man, was obsessed with it. He championed the "Theory of Recapitulation," which stated that during the time it takes to develop, an embryo passes through all the forms of its ancestors, in a rapid, compressed manner—rather in the way that the life of a drowning person flashes through his or her mind. If this is true—and recently the theory has lost much credence—the study of the development of any embryo can be used to explain much about its ancestry. A human egg, for example, would start off with primitive fish-like characteristics, and throughout the nine months of gestation would resemble, in turn, an amphibian, a reptile, and a monkey, before taking on the sophisticated features of a human.

Haeckel, who was known formally as His Excellency, Herr Geheimerat Professor Dr. Ernst von Haeckel, was completely convinced by it. For decades, all European schoolchildren were made to repeat his famous formulation, "Ontogeny recapitulates phylogeny"—the development of the individual mimics the development of the species—until it was etched indelibly on their brains. In 1891 he dispatched one of his disciples, Professor Richard Semon, to Australia to look into the ontogeny of *Neoceratodus*.

Semon returned to Germany two years later with seven hundred bottles, each containing a pickled lungfish at a different stage of development—from egg to fully formed junior lungfish. He reported his inspection in an exhaustive paper. He had learned, among other things, that *Neoceratodus* does not estivate (entomb itself in mud during the summer) like its African and South American cousins, and that its single lung cannot keep it alive on dry

land. He also discovered, however, that as long as there is sufficient moisture, it has—rather like its cousins, *Lepidosiren* and *Protopterus*—internal nostrils that help to keep it going during the dry
season, when other fishes perish. Even when in plentiful freshwater, he observed, the fish came to the surface for a gulp of air
about every hour.

Of most interest to the missing linkists, however, was the
result of his scrutiny of the lungfish's strange, limb-like fins: while
Neoceratodus often used them like legs in the water, they were not
strong enough to support it on land. Semon's most important conclusion of his work into the ontology of the Australian lungfish was
that, contrary to what previously had been believed, it was not a
direct ancestor of amphibians, but was more like a close cousin.
The race to identify the first walking fish was still on, and the
coelacanth was back in the running.

Since the fossils of all of these creatures came to light in the nineteenth century, scientists have been debating the interrelationships
among the different lobe-fins, and fighting for their particular
favorite to be acclaimed as the true ancestor of man. For a while,
in the first half of the twentieth century, the coelacanth was flavor
of the month; then the accolade passed back to the lungfish, on to
the Rhipidistia—now extinct—and back again. Even today, more
than sixty years after the discovery of the first coelacanth, there is
still no clear conclusion.

Determining the interrelationships of creatures is a complex
process, and scientists are forever inventing new ways of doing it.
In his book, *Living Fossil: The Story of the Coelacanth*, biologist
Keith Thomson describes a conundrum put before a meeting of
zoologists in Reading, England. Among the following three creatures—the salmon, the lungfish, and the cow—they were asked,
which two are most closely related? (They might easily have used
the coelacanth instead of the lungfish.) The answer to nonscientists
might seem immediately obvious: the two fishes. But not to zoologists. To them it is clear that it is the lungfish and the cow,

because they belong on a similar line of descent and thus have a more recent common ancestor: the lungfish is at least a cousin of tetrapods (albeit many times removed), while the salmon, whose line branched off much earlier, is only a cousin of the whole group. Thomson uses an analogy: who are most closely related out of Queen Elizabeth II, Kaiser Wilhelm II, and Margaret Thatcher? Although the queen and the former prime minister are both female, British, late twentieth century rulers of countries, the queen clearly is more closely related to the Kaiser, since both are descendants of Queen Victoria. Thomson's point illustrates that in deciphering relationships between organisms, one must not look merely at appearances, but at ancestry and the clues provided by uniquely shared characteristics. But the characteristics of the contenders are complex and often conflicting.

An examination of the characteristics shared by the living coelacanth, the Australian lungfish, a living amphibian (say, a newt), and a common ray-finned fish (such as a salmon) showed clearly that the lungfish had most in common with the newt, while the ray-fin had least. The coelacanth was somewhere in the middle—exhibiting enough newt-like characteristics, like its limbs, inner ear, and gills, to show that a relationship existed, but not enough to indicate it was a very close one. The forefather of mankind, it appeared, did not look exactly like *Latimeria chalumnae*.

Then again, perhaps our great-great-grandfather many millions of times removed did not look too different from the lobe-finned fishes of the Devonian era. The fossil coelacanths, the now extinct rhipidistians, and the primordial lungfish shared significant characteristics—with each other, and with the fossils of early amphibians such as Ichthyostega. Of particular notice was the structure of the head casing and paired fins. The lungfish had very similar fringed tails, and from fossil reconstructions, one can see that a rhipidistian looked like a longer, thinner, smaller version of today's coelacanth, with the same strong notochord and pudgy fins. Fossil evidence of the Rhipidistia died out in the early Permian period, around 300 million years ago (although, given the surprise "reappearance" of the coelacanth, which also was thought

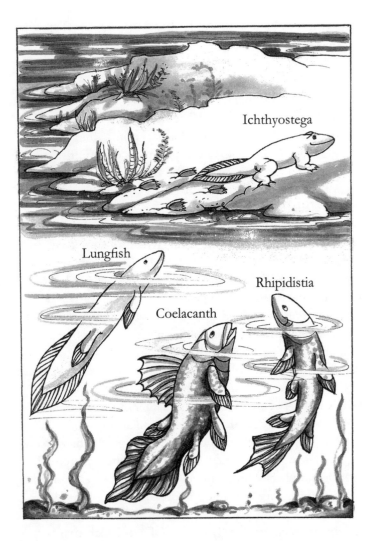

A question of prehistoric paternity: 400 million years ago,
one of these three fish crawled out of the sea to conquer the land.

to have been extinct for many millions of years, one has to be careful in signing its formal death certificate).

The only sure thing is that debate over which one of the

Sarcopterygii gave rise to the first fish to crawl out of the sea will run and run. The latest theory is that, while the lungfish is a "sister" of the first tetrapod, the coelacanth is a "cousin"; but again, this has not been proved. Even direct comparisons of the genetic codes of the modern descendants through DNA analysis have been inconclusive. Recent results show that coelacanth chromosomes are remarkably similar to those of primitive frogs, and that the lobe-fins and tetrapods are clearly related; however, within the group, none shows up as being more closely related than the others.

DNA testing was not available to Millot and Anthony in the 1950s and 1960s. However much they learned from their investigations, much about the coelacanth remained a mystery. Quite apart from such complex issues as where exactly the coelacanth fit into the evolutionary tree, they still couldn't answer more fundamental questions about the fish, like how did it reproduce. One fossil showed the vague imprints of tiny babies inside the outline of the adult. This was read as evidence that coelacanths gave birth to live young—fully functioning little coelacanths rather than eggs. (This indicates a high level of sophistication—only 5 percent of fishes are, like mammals, live-bearers. A female cod, for instance, lays a million eggs, which are swept away by the current, and relatively few of which get fertilized; while lungfish, like most amphibians, lay large, yolky eggs into nests, which they protect from predators.) In 1955, and again five years later, this theory apparently was thrown out of the water when the first female coelacanths were dissected by Millot and were found—again, like sharks—to carry immense, delicate eggs the size of grapefruits, the largest known fish eggs. Perhaps the coelacanth was not, after all, ovoviviparous (live-bearing), but oviparous (egg-bearing)?

After years of detailed study, Millot and Anthony had gathered an enormous amount of information on the coelacanth. Rather than supplying simple answers, however, their work raised more questions. The coelacanth was not simply a key to the past, and the door to its ancestry could not immediately be unlocked by an examination of its anatomy. It was clear that, in *Latimeria*, they had a fish with a unique combination of characteristics, some of them in common with amphibians, some with fish; and which, over millions of years, had found different solutions to the problems posed by evolution.

When Millot and his colleagues opened the time capsule and revealed to the world *Latimeria*'s inner secrets, they did not find a blueprint for our ancient ancestor, but instead a confusing potpourri of characteristics. They soon came to accept that they were dealing not with an old fossil, but with a contemporary fish, which had adapted to survive for 400 million years while the world around it—the oceans and the continents—had changed beyond all recognition. Few of its fellow creatures had been around for much longer than a twinkling in the coelacanth's eye; the Indian Ocean had not even come into existence until more than 100 million years after the coelacanth first swam in primordial waters. As interesting as what the fish shared with ancient creatures was how and why it had survived the vicissitudes of time and nature, the dangers of fierce predators, when so many other creatures had perished—why it had, in a sense, trumped evolution and endured to become a living fossil.

The fish on the slab in front of Millot was important as more than just an obscure link to our distant past.

VII

LE POISSON VIVANT

Fairly early on in his investigations, Millot had realized that if he was to make any major progress in understanding why and how the coelacanth had survived for so long while the vast majority of creatures had become extinct, he would have to see a coelacanth in its natural environment. It was one thing to note the limb-like fins or study the muscles—but it would be quite another to see how they were used. It was imperative for science to get hold of a live coelacanth.

This would not be easy. In 1954, the famous Jacques Cousteau was summoned to the islands in his equally well known ship, *Calypso*, to try out his pioneering deep-sea diving methods in Comoran waters. Under Millot's direction, he took reels of submarine photographs in areas thought to be possible coelacanth lairs. But on this occasion, even Cousteau's legendary Gallic charm failed to entice the coelacanth from its ocean hiding place.*

Millot turned his attention to the local fishermen. Comorans fish from their rough wooden *galawa* canoes. Each *galawa* is not

---

*Cousteau tried again in 1963 and 1968—and again he failed to find a coelacanth to film.

much larger than a coelacanth, and hitherto, once a fisherman had caught a coelacanth, he would immediately batter it to death or slit its throat to prevent it from struggling, and then haul it into the boat. The fishermen were afraid that if they left their valuable catch in the water and attempted to tow it to land behind the boat, a shark or barracuda might grab it, and they would lose both the fish and their reward. So in order to make the capture of a live coelacanth more attractive, Millot doubled the reward to £200. He also set about devising a suitable watery jail for the captured fish.

The large reward was enough of an incentive for the fishermen, even if they were not altogether successful. Early one morning in 1954, Dr. Guy Arzel, who was at that time in charge of the simple hospital on Grande Comore, was awoken by a pounding on his door. He opened it to find a fisherman cradling a bleeding coelacanth in his arms. He pleaded with the doctor to save its life. He had caught the fish only a short time ago, he explained, and when he saw it appearing out of the blue depths and realized that it was a *gombessa*, he was determined to bring it to shore alive and claim the large reward. As the fish broke the surface, however, his courage failed him, and in a fit of terror, he bludgeoned the fish repeatedly on its head. He had brought it to the hospital in the hope that Dr. Arzel could somehow bring it back to life. Unfortunately, this was beyond the powers of the French doctor, who placed the fish on his operating table, but not knowing how to proceed, had to stand by and watch the beautiful blue fish before him slowly fade and die.

Shortly after this, an Anjouannais fisherman had better luck. On November 12, 1954, at 8 P.M., Zema ben Said caught the eighth coelacanth, 1,000 meters off Mutsamudu, on the northwestern coast of Anjouan. It was two days after the full moon, and the sea was calm. From the way the fish took the bait, he guessed it was a *gombessa*, so he took his time, carefully hauling the fish in by hand from a depth of 140 fathoms (255 meters). Having made sure it was indeed *"Le Poisson,"* Zema decided to try for the doubled reward. He gently passed a cord through its mouth and out of the gill opening, and thus he led the creature back to Mutsamudu jetty; though sometimes it was the fish that towed the *galawa*.

The administrator of the island was notified immediately, and as planned, sunk a seven-meter whaler off the end of the jetty, leaving the plug out of the bottom to allow a small current to flow through. By 9:30 P.M., the makeshift aquarium was ready. The live coelacanth was released into it, and the boat was covered by a net to prevent the fish from escaping. Every half an hour, the boat was rocked to enable freshwater to enter. Witnesses reported a dark grayish-blue fish, the color of watch-spring steel, with luminescent greeny-yellow eyes.

That night, the population of Mutsamudu celebrated the valuable catch. They sang and danced until daybreak, making regular trips to view their precious fish. The coelacanth at first appeared bewildered but otherwise calm; it swam slowly, by curious rotating movements of its pectoral fins, using its second dorsal fin, anal fin, and tail as a rudder. As the sun rose, however, it became increasingly distressed, apparently reacting against the light and the heat, trying to hide in the darkest corners of its makeshift cage. Several tent canvases were placed over the boat in an attempt to protect the fish from the sun, but to no avail. By 3:30 P.M., it was lying belly-up, its fins and gill covers making agonized movements.

Millot arrived from Tananarive on a special flight just in time to see the coelacanth take its last breaths. He hauled it from the water, covered it with a sheet, and took it to the hospital, where it was formally pronounced dead—but in perfect condition for scientific examination. Zema received his full reward.

"There can be no doubt that death was brought about by decompression combined with rise in temperature. . . . " Millot wrote in his account of the event, published in *Nature* in February 1955. "It must also be noted that the *Latimeria*, which appeared greatly distressed on its arrival at the surface, seemed to have recovered appreciably after an hour or so and passed the rest of the night quite comfortably without any too obvious inconvenience. It was daybreak, either the appearance of sunlight or the gradual warming of the water, which initiated the progressive discomfort which led to its death." Millot proposed that next time the experiment was attempted, the coelacanth should be incarcerated at a depth of 150 to 200 meters, and hauled up only for observation or to be photographed.

Unfortunately, there wasn't a next time. During Millot's reign over the Comoran coelacanths, repeated efforts to capture or film a live coelacanth failed, and Zema ben Said was the only Comoran fisherman to claim the award of £200.

Four years after his dramatic flight to the Comoros on December 28, 1952, J.L.B. Smith had been out of the coelacanth picture for some time. While his examination of *Latimeria* had been remarkably thorough as far as it went, by virtue of the missing soft parts, he had been unable to make much of a contribution to knowledge of the fish's internal workings. After the French got a specimen of their own, Smith left the close anatomical work to Millot and his team, and *Malania* was put on display at the Rhodes Department of Ichthyology.

He never ceased to feel he had some sort of dominion over the creature, however; he had, after all, introduced it to the world. He launched a vehement attack on what he saw as French profligacy with regard to the coelacanth. He used that most august of arenas, the letters page of the *The Times* of London, on June 4, 1956.

Sir,—

> *The tragic presumed loss of Captain Eric Hunt in shipwreck not far from where he achieved fame in 1952 by reporting the second coelacanth at the French Comoro Islands brings forcibly to mind that it is now well over three years since this unique ichthyological survivor was shown to live there. Owing to circumstances it was at that time necessary to seek complete specimens for scientific purposes, and it will be remembered that not long after this discovery the French prohibited foreign scientists from searching for coelacanths in their waters, proposing instead that this should be done by an international expedition under French direction. . . .*

> *The Comoros are of unique structure, with steep slopes that plunge abruptly to great depths in the ocean. . . . All the evidence*

indeed indicates that there can not be any high population of coelacanths there. If, as is quite likely, these are the only coelacanths in existence, there may well be only a few hundred of them in all. In the past three years the French have got ten specimens, more than enough for full scientific examination. Science and the world are no longer crying for dead coelacanths. . . .

With my fundamental interest in this remarkable creature I have become increasingly uneasy about the situation affecting coelacanths. . . . If a herd of dinosaurs were discovered in some remote jungle the world would rightly recoil in horror from a policy of rewarding natives to slaughter as many as possible. The situation of the Comoran coelacanths is in reality no better, and the present policy is debasing a once important scientific quest to the level of senseless slaughter of one of our most precious heritages in biology.

The French own the Comoros, but the coelacanth belongs to science and to mankind. French scientific authorities carry the gravest responsibility in this matter. The policy of rewarding natives for catching coelacanths should immediately be reversed, or modified, to that of a severe punishment for killing one. . . .

I am, Sir, yours, &c.,

J.L.B. Smith

The response was rapid and indignant. Gavin De Beer, director of the British Natural History Museum, was the first to weigh in. The following day, his response was published in *The Times*.

Professor J.L.B. Smith's letter today implies criticism of French scientific methods which is not deserved. When he says that 10 specimens are more than enough for scientific investigation he appears to forget that only two of these are female, neither of which is in a condition to show whether the fish lay eggs or give birth to live young. . . .

When Professor Smith states that science and the world are no longer crying for dead coelacanths he forgets that, while he may possess a specimen, other museums such as this have need for one, and have been promised specimens by our generous French colleagues.

The next week, Millot, the main target of Smith's attack, wrote in his own defense to *The Times*, managing at the same time to get in a few, possibly deserved, below-the-belt blows at the South African.

> *If Professor Smith were an anatomist he would realise that, for an extensive study, anatomical, histological, and chemical, of the kind which we are pursuing on coelacanths of both sexes and all stages, a dozen specimens (which we do not even yet possess) are hardly sufficient. . . . I am in a position to give your readers a full guarantee that not one coelacanth caught has been superfluous. . . .*

He paid tribute to the cooperation of the Comoran authorities before directing one last jab in Smith's direction:

> *In particular they [the Comorans] must be credited with having protected the coelacanths from being dynamited with depth-charges, which, though hardly believable but nevertheless true, Professor Smith twice proposed to do.*

By 1960, Millot had decided that he had enough dead coelacanths of his own, and started to give spare ones away. The British Natural History Museum was rewarded for its loyalty in *The Times* battle against Smith with Specimen 14; Specimen 21 went to the Zoological Museum in Copenhagen. All of these were presented on the understanding that they were for exhibition purposes only, and not to be dissected for research.

After a few more years, Millot apparently tired of even his own generosity. In 1962, Dr. Georges Garrouste, who had been involved in the battle to keep Zema ben Said's fish alive, wrote to J.L.B. Smith complaining that, three months after the event, the Institut de Réchèrche Scientifique at Tananarive had yet to acknowledge receipt of a coelacanth he had sent them. "In spite of our requests [they] have not yet rewarded the fisherman. In short, as a result of this negative attitude and apparent lack of interest, it has occurred to me that other scientists might be happy to benefit from such a capture." He went on to offer Smith Specimen 26, a massive female, 5 feet 11 inches (1.8 meters) long, weighing a hefty 209 pounds (95 kilograms). J.L.B. declined, however, and suggested the fish be offered to the American Museum of Natural History, who gratefully accepted. There was little more he could add to the detailed work already being undertaken by the French, he felt. According to the ever loyal Margaret Smith, "He always maintained that the first coelacanth had given him more than any man could hope for in a lifetime."

The coelacanth had occupied the first place in Smith's thoughts for nearly two decades. With his refusal of a fresh specimen came an acceptance that it was no longer his fish, but rather *"Notre Coelacanthe."* Typically, however, he did not slow down. *Old Fourlegs: The Story of the Coelacanth* was published in 1956. William Smith maintains that J.L.B. went out fishing one day in Knysna, then returned home with the whole book in his head and proceeded to write it in ten days flat. It was published in London and New York (under the title *The Search Beneath the Sea)* and translated into German, French, Russian, Estonian, Afrikaans, Slovak, and Dutch. J.L.B. dedicated the book to "Miss M. Courtenay-Latimer, one of South Africa's most able women." Its opening sentence is characteristic of Smith, a scientist to his core: "These are wonderful times, and it is thrilling to be living now, though it would thrill me even more to know that I could still be here a hundred or a thousand years hence, for this immediate future promises to be of intense interest, even excitement, certainly to the scientist."

The Smiths continued working on new editions of the *Sea Fishes* book, and after an expedition to the Seychelles, this time accompanied by William, they published *The Fishes of the Seychelles.* After 1957, J.L.B. no longer undertook overseas expeditions, although he continued to travel to conferences abroad until 1960, when out of consideration for his always fragile health, he stopped traveling altogether and dedicated himself to his work, and to terrorizing his staff and pupils.

J.L.B. undoubtedly had a softer side, but as he got older it was allowed to show less frequently and his reputation for intolerance blossomed. Shirley Bell, at that time a young writer working for an angling magazine, was among the honored few whom he always had time for, and the J.L.B. she remembers is a very different man from Peter Barnett's stern professor. She received a letter from the Smiths praising one of her articles. "That began our friendship," she remembered. "He wrote me long letters in his neat spidery handwriting . . . discussions about his articles that I was publishing in the magazine, news of their current activities in ichthyology, bits of advice, incisive comments in reply to my questions, gentle concern . . . wonderful letters. . . .

"Nothing was too much trouble for him," she explained. "He had an instinct for thoroughness and he noticed the smallest thing that he considered had been well done. He had a reputation for remoteness and even for eccentricity, and he certainly didn't suffer fools gladly, but that was not the side of him I saw in the few years of our friendship."

Smith also remained close to Marjorie Courtenay-Latimer; his archives are full of typed carbon copies of warm letters to her, mainly, inevitably, about fish. Even after thirty years of friendship, he addressed her as "Miss Courtenay-Latimer," while she replied to "Prof. J.L.B." His other great companion and fellow fish lover, a fox terrier called Marlin, was born in 1959. They were seldom apart. Marlin accompanied J.L.B. wherever he went—in the car, in the boat, on long walks around Grahamstown.* Marlin even accompanied the Smiths on a trip to Marjorie Courtenay-Latimer's Bird Island in 1964, a trip he clearly enjoyed, as J.L.B. explained in a thank-you letter to the lighthouse keeper and his wife soon after their return. "We have been to many unusual places in the course of our work, but I do not think we shall ever forget Bird Island. We had a not unpleasant trip back to the mainland, Marlin, I think, suf-

---

*In 1993, on the very route J.L.B. and Marlin used to tramp daily, scientists from Rhodes University unearthed the fossil remains of seven coelacanths among a multitude of other fish fossils on the site of a prehistoric lagoon, only two miles from Grahamstown.

fered most, he hated the motion of the boat. This was probably due to the fact that he had thoroughly exhausted himself chasing rabbits, indeed his feet were sore for days afterwards."

J.L.B. devoted the last years of his life to the fish books and to the Department of Ichthyology. His determination never flagged: from the time he first became interested in ichthyology, he published over 500 papers on fish, and named 370 new species. Four honored his wife and partner, Margaret. In dedicating *Pseudocheilinus margaretae* to her, perhaps untypically he allowed his private emotions to show: "This exceptionally beautiful creature is named as a small tribute to my wife," he wrote, "whose contribution to all phases of our work is probably greater than my own."

Honors were showered on J.L.B. Smith. He was made a Fellow of the Royal Society of South Africa, an Honorary Foreign Member of the American Society of Ichthyologists and Herpetologists, and a Corresponding Foreign Member of the Zoological Society, London, among others. But latterly, he refused some of the honors offered to him. "Give it to a younger man who is still climbing and who would appreciate it—it would be wasted on me," he said in response to one. He agreed, however, to receive the degree of Doctor of Science *honoris causa* from Rhodes University in April 1968.

For the last two years of his life, J.L.B. felt his mental powers were deteriorating. Margaret was taking on an increasing burden of her husband's work at the Department of Ichthyology. His eyesight was failing and he was terrified of having a stroke. One of his greatest university friends, the former minister of internal affairs, Eben Donges, had died after a stroke a couple of years earlier. Smith, always conscious of his ailments, dreaded becoming bedridden and a "useless hulk."

At the end of November 1967, he handed his secretary, Jean Pote, a large check. It was twice as large as the Christmas bonus she had expected. "I want you to keep it," he told her. At Christmas, he kissed his daughter-in-law Gerd, Bob's wife, on the forehead. "I was surprised. It was the first time he had shown any sign of affection," she recalls. He was similarly demonstrative when he saw his old friend and coelacanth compatriot, Marjorie Courtenay-Latimer, late that year.

"Whenever he came to East London, he would have lunch with me. So this time, we had eaten, and when I went out to the car to say good-bye, he threw his arms around my neck and kissed me on the cheek. I thought it was very strange—he had never done that before. Then he said: 'Well, lass, you're doing a fine job, carry on.' I laughed and thought what a funny thing it was for him to say."

On January 8, 1968, aged seventy, at his home in Grahamstown, J.L.B. Smith took a fatal dose of cyanide. It was carefully planned and neatly carried out. He left two notes.* The first was addressed to Margaret: "Goodbye my love, and thank you for a wonderful thirty years. I am going upstairs to the servant's room. Careful. Cyanide."

In the second, he explained himself. He had typed: ". . . For some years I have suffered from severe mental depression . . . the sight of one eye has almost gone . . . back pressure is proving troublesome . . . I live in perpetual fear of becoming bedridden and helpless. . . . I prefer to take this way out, probably only a brief anticipation of nature."

J.L.B. Smith was mourned across the world by scientists, anglers, and friends. Margaret received a flood of tributes:

"I have always treasured our friendship and I have especially appreciated his kind attention and very wise advice he so readily gave me," wrote the scientist R. Liversidge. "I owe him a lot and far more than most realised . . . for the philosophical attitude he held towards science, administrators and his fellow men."

Humphrey Greenwood of the British Museum described how "as a student I stood in awe at the work he did and had done. Now, having had many years' experience of similar research, the awe is further emphasised."

At the opening of an exhibition at the East London Museum to celebrate J.L.B. Smith's life and work, the president of the Council for Scientific and Industrial Research, Dr. H. J. van Eck, compared the Smiths' scientific collaboration to the Curies' and the Webbs'.

"I see him now as I always knew him, lean, virile, fearless

---

*Both of these notes have disappeared from the Smith files.

J.L.B. Smith with Marlin at Knysna, shortly
before his death in 1968.

and absolutely honest in all his dealings with both men and with
scientific facts and theories," lamented a friend, Dinnie Nell.

"I would have turned to him before anybody else amongst
my friends and acquaintances and been quite sure of warm sympa-
thy, advice and any needful assistance," the Ichthyology Department librarian Doris Cave wrote.

His old friend and coelacanth companion, Marjorie Courtenay-
Latimer, was overwrought. "How honoured we were to know him,"
she wrote to Margaret Smith. "He was a brilliantly clever man, and I
was very, very lucky to have known him."

VIII

IN COMORIAN

By 1963, the French had decided they had enough specimens to work with. They moved their base of operations to a special laboratory at the Muséum National d'Histoire Naturelle in Paris, and handed over the distribution of further fish to the Comoran authorities. The Comorans were delighted, and instituted a practice whereby the fishermen had to sell coelacanths to them for a fixed reward of £100, and they would, in turn, dispose of the fish how they wished. Picture pamphlets were distributed to all the schools in the Comoros, explaining what the coelacanth was, and urging the children to go home and tell their parents. The sole Air Comoros plane was put at the disposal of the fish: whenever one was caught, the plane would be dispatched immediately to bring it back to the government freezer in Moroni. The French still monitored the catches, and for the next decade, until the Comoros declared their independence and the French left, records show an increased catch. Nine coelacanths were fished in 1965, and six the following year.

It is a testament to the remoteness of the Comoros that before 1952 the coelacanth was not known to live there. Legend has it that King Solomon married the Queen of Sheba on the remote Indian Ocean islands a thousand years before the birth of Christ, but evidence of a more archeological nature dates the arrival of the first settlers during the first century A.D. They sailed from Indonesia, six thousand miles across the Indian Ocean in narrow dugout canoes, which were several times longer than the *galawa* used by fishermen today but similar in shape, with the same double outriggers. They were men of the sea, and when they colonized the islands of the western Indian Ocean—first Madagascar and then the Comoros—they brought with them their skills and techniques.

They found four perfect jewels strung across the neck of the Mozambique Channel. The volcanic soil was fertile, and the sea full of riches. It is not known when they first caught the strange fish they called *gombessa*, but it is possible they have been fishing it up for hundreds, perhaps even thousands of years. Some of the time, they just threw it away. It was *nessa*, the oil fish (*Ruvettus pretiosus*), they wanted; *nessa*—similar in size and habitat to *gombessa*, the coelacanth—commanded a good price at the market, and still does to this day. It is bought for its medicinal qualities— its oil-rich flesh acts as both a laxative and a mosquito repellent. *Gombessa*, on the other hand—which in Swahili translates as something taboo, strictly forbidden—does not taste good, and its oily flesh is a violent purgative. Its value lies in its ancient pedigree, which the Comorans neither knew nor cared about.

Over the centuries, the first Comorans were joined by Arabs, Shirazis, and African immigrants; today, you can clearly see the mixed ancestry in the Comorans you meet. Soon after the birth of Muhammad in A.D. 570, a Comoran emissary returned from Arabia with a caliph, who converted the country to Islam.

By the fifteenth century, the islands were a center of maritime trade, ruled by power-hungry sultans constantly at war with one another. Ships plying the wealthy east African coast stopped off in the Comoros to trade rice, ambergris, and slaves. Inevitably, they attracted the attention of pirates, who used to hide in secret coves, ready to pounce on the overladen trading vessels. By the beginning of the last century, however, the Comoros had subsided into civil

The Friday Mosque in Moroni, Comoros.

war, as the sultans battled with one another to supply slave labor to eager French plantation owners on Ile de France (now Mauritius) and the Seychelles.

The end of the slave trade and the opening of the Suez Canal in 1869 effectively banished the Comoros to a backseat in the theater of international trade. By the close of the century, only two or three ships stopped at the islands each year, and the country slipped into penury. France stepped in, and in 1946 the Comoros became a formal colony. French loaves and *pains au chocolat* were integrated into the Comoran diet of fish, rice, and cassava, and even today, on every street corner in the capital, Moroni, you can find a Comoran lady sitting in front of her bag of crusty baguettes.

The Comoros were among the smallest and most remote of French colonies. Even the French foreign office probably had little idea of their existence before attention was sharply focused on them after the diplomatic debacle of Smith's appropriation of the 1952 coelacanth. Les Comores were nothing but a small drip, drip from Quai d'Orsay coffers. When the discovery of *Malania* was announced to the world, it was not only J.L.B. who reached for an atlas.

By the second half of the twentieth century, the islands were home to half a million people, mostly living in small coastal villages. Around each village were pockets of subsistence farming: vanilla, fruit, and spices sharing the volcanic slopes with small herds of cattle and goats. There was no industry, few raw materials to export, and little for the men to do except fish, play football and discuss local politics.

The coelacanth provided a boost to both the Comoran pride and the economy. Sidi Bacari Papa, the taxidermist on Anjouan, was kept busy, as favored visiting heads of state and other dignitaries were presented with a stuffed or frozen coelacanth: the United Nations received one, as did President Mitterrand and South African foreign affairs minister "Pik" Botha. It was a gift no other nation had the power to bestow. The coelacanth was proudly minted onto Comoran coins and notes, it was printed on colorful pareos and T-shirts, and fashioned into delicate gold brooches and necklaces, bridal *grand mariage* presents. In December 1986, the minister of the interior declared it to be a "heritage of humankind," and the Comoran nation its warden.

The fishermen who caught a coelacanth benefited, too. The islands are extremely poor, and for most people, subsistence farm-

**The coelacanth as a Comoran cultural symbol:
on coins and banknotes.**

ing and fishing is a way of life. Few people starve, but fewer still have money to burn. The society is strictly stratified: at the top of the pile sit the sharifs, the descendants of the prophet Muhammad. Below them are the intellectuals, the *notables* of the mosque, the civil servants, and the teachers. Then come the farmers, and below them, on the very lowest rung of Comoran society, perch the fishermen. They are regarded as coarse—loud-voiced and fond of a brawl. On the whole, the sons and grandsons of fishermen themselves become fishermen and marry the daughters of fishermen. Even if they work hard to become intellectuals, they are still marginalized by society. There is a Comoran saying: "Even if you make fish soup, it still smells of fish."

For a while, the coelacanth changed all this. Suddenly, the fisherman who caught one was a local hero, feted by the *mzungus*, white men who hovered around the fishing villages, urging them to hook a *gombessa* and eager to find out all the details when they did. Apart from Captain Goosen's chance trawl in 1938, every coelacanth had been caught by Comoran fishermen using traditional techniques. The fisherman who sold a coelacanth also suddenly had more money than most people earned in five years. Even in the Comoros today, £100 is a lot of money, a jump start on the route to a *grand mariage*, the cornerstone of Comoran society and the method by which an ordinary man of the village becomes a respected *notable*.

Mzé Lamali Hila started fishing when he was ten years old. He claims to be over a hundred now, and still climbs from his sagging rattan bed into his *galawa* for nightly fishing trips. These days he doesn't catch so much. "I have caught four *gombessa*," he says with pride. "And I remember when I was a small boy my father caught one. It doesn't taste good,* but we used it as medicine for many diseases; upset stomachs, spots, hip problems. The first one I caught was a long time ago," he recalls. A small crowd gathers at the entrance to Mzé's ragged palm weave hut, and the old fisherman starts playing to his audience, telling his story in a melodic, swooping voice. "I was using a method we call *mazé*, which we use for the deep-sea fish, *nessa* and *gombessa*. I attached two flat stones to my line, about this far [he indicated the length of a forearm] from the hook. For bait, I used a small fish, *roudi*, which *gombessa* finds very delicious. I let the line out and when I sensed the ocean bottom, I gave a sharp tug and released the stones. Very soon I felt a bite. I started pulling it up to my boat, very slowly; it took almost two hours and it was very heavy. When I got it to the surface, I put a very big hook through its mouth to attach it to the boat. It made a wave as I pulled it to shore. In those days, before the whites came, it had no value—you couldn't sell it at the market, so the next day, we cooked it. It was not delicious, the flesh is too fatty, not good, and many people didn't eat it—only some children and villagers from up the hill. If you eat *gombessa* or *nessa*, you have to have a shower," he recounts to the amusement of his

---

*On March 24, 1975, the East London *Daily Dispatch* reported that California scientists had dined on coelacanth fillets. They had defrosted some frozen muscle, cooked and eaten it. According to Dr. John McCosker, "It was absolutely delicious."

Comoran fishermen at sunset.

audience. "It gives you diarrhea." After 1952, he fared much better. He sold his second coelacanth for 100,000 CFA (almost $300); his third and fourth for 50,000 CFA and 40,000 CFA respectively. With rewards like that, every fisherman was eager to hook a *gombessa*. While there is no proof that they can fish them on demand, the monetary incentive enticed more fishermen to go out at night, leading inevitably to a bigger catch.

The Comoran government was doing a roaring trade in coelacanths. In 1966, biologist Keith Thomson was working at the Peabody Museum of Natural History at Yale University when they received a circular letter from the Comoran government, offering any zoological institution the opportunity to purchase its own coelacanth for the bargain basement price of $400, plus shipping. The demand, inevitably, was high, and today a coelacanth occupies a place of honor in many of the world's premier natural history museums, from Los Angeles to Tokyo. The Peabody managed to get hold of the first fresh-frozen coelacanth, which was displayed, prior to dissection, in an ice-cream freezer case. "Most of Connecticut filed past," Thomson reported. "It was a miniature version of what J.L.B. Smith must have endured. The Great Hall of Yale's Peabody Museum came to resemble nothing less than Lenin's Tomb."

Once the French had left the Comoros for good in 1972, permission was granted for overseas expeditions to hunt for their own coelacanths around the islands. What they all wanted was a live specimen; apart from a few dying fish paddling around in the shallows, no one had seen a living coelacanth.

The potential rewards for the first images of a live coelacanth were so great that photographers went to dramatic lengths to obtain them. Only a month after the capture of *"Notre Coelacanthe"* in 1953, an Italian zoological expedition led by Franco Prosperi claimed to have photographed a live coelacanth off Mayotte. In his book, *Vanished Continent,* Prosperi tells of how he leaned over the side of the dinghy and peered into the shallow water. "For some minutes a strange fish had been occupying my attention. It was resting on a madreporic [hard coral] rock about forty feet below the surface. As I say, it was not swimming, but was lying on the corals, too lazy even to move a fin. . . . I carefully studied its squat body, its elevated trunk, and its uniformly dark colouring." He recognized it by its strange fins to be a coelacanth, and with his "heart pounding," dived into the water and swam towards the putative coelacanth. "I saw the outline in the viewfinder," he wrote. "I noted the sturdy appearance of the body; the fins protruding like small spatules from the end of the fleshy peduncles." He took a picture, but sadly for him, "The sudden click and the subsequent whirring of the mechanism galvanised the fish into life. Turning with a speed incredible in such a ponderous creature, it headed rapidly for the sea bottom. While I was trying to follow it I saw Fabrizio swim past on my right, his body rigid behind his gun."

Fortunately for the coelacanth, Fabrizio didn't manage to get a clear shot. Soon afterwards, the French issued their edict forbidding foreign scientists from catching a coelacanth, and the coelacanth population was spared the Italians and their underwater armory. The single photograph of the coelacanth was printed in *Vanished Continent,* with extravagant claims of it having been the first "living" picture. Scientific fraud antennae immediately started to twitch. The picture is unclear, and there are various other indications that the encounter was a fake. For a start, another coelacanth has never been seen off Mayotte, or in such shallow water; a live coelacanth is not of a uniformly dark color, but covered in white splotches; and it has never been observed resting on or even touching the surface of the coral. Most coelacanth experts believe the picture was somehow falsified.

Thirteen years later, a French photojournalist, Jacques Stevens, sold a story to *Life* magazine in which he purported to have filmed a coelacanth swimming in the "dim and aerie" depths of the ocean. He was diving at night when "out of the spooky darkness at about 130 feet below the surface, here came a coelacanth." The fish was "mucus-covered," and its "huge phosphorescent eyes glared at me." Unfortunately for Stevens—and posterity—his movie camera jammed, and the flash of his still camera appeared to frighten the coelacanth, which, he claimed, swam off after he had taken only two pictures.

When the story appeared, scientists again quickly recognized that all was not exactly as it was intended to appear. The fish was photographed against coral that is found only in shallow water, and the pictures themselves were well lit and clearly taken in sunlight.

In addition, they saw marks on the coelacanth's snout consistent with the effect produced by the rubbing of a fishing line; and it had cloudy eyes and exhibited clear signs of stress. It was apparent to the scientists that this denizen of the deep had, in fact, been caught by a fisherman in the usual manner, then taken to the shallows, where Stevens had photographed it in its dying stages.

Another Frenchman, local dive master Jean-Louis Geraud, made no bones about the coelacanth he filmed being on its last legs. "It was wonderful," he said in heavily accented English. "It was exactly like I see a dinosaur in my garden, but a nice dinosaur because he swim like he dance."

In 1972, a joint Franco-Anglo-American expedition failed to catch their own fish, but were on the scene when two coelacanths were caught by local fisherman. The first, a large female, was dissected on the spot and found to contain nineteen huge eggs, the size of grapefruits, weighing between 6 and 8 pounds (300 to 350 grams). These were the largest fish eggs in the world, and further suggested that the coelacanth gave birth to eggs instead of live young.

But in 1975—thirty-six years after J.L.B. Smith examined his first coelacanth—that theory was finally knocked on the head. The Comoros, under the leadership of Ahmed Abdallah, had declared unilateral independence from France, and the French embargo against international scientific institutions examining their specimens no longer held. The American Museum of Natural History immediately started their dissection of Specimen 26, the vast coelacanth that originally had been offered by Dr. Georges Garrouste to J.L.B. Smith, but which he had declined. When they opened up the fish, they were amazed to find five almost fully developed *Latimeria* pups, close to 30 centimeters long, and each attached to the remnants of a large yolk sac. The fossil evidence had at last been proved: *Latimeria* was a live-bearer.

This knowledge had important implications: if the early coelacanths had also borne live young, in doing so they would have predated the first mammals by 100 million years. But at the same time, it indicated a slow population-growth rate. The coelacanth is believed to have a gestation period of over a year; if it produces only five young, then the rate of renewal of a small coelacanth population is necessarily slow, and any increase in predation might wipe them out completely in a short period of time. A group of scientists, led by Eugene Balon of Guelph University in Canada, went further and surmised that the small number of embryos—versus the large number of eggs—was evidence of oophagy, or uterine cannibalism, which is also common in sharks. This theory was immediately refuted by another group of coelacanth specialists—but the truth was that none of them knew for certain. Calls increased for the capture of a live coelacanth which, it was thought, would enable such mysteries to be studied closely.

The Steinhart Aquarium of San Francisco raised the stakes by offering the successful fisherman not only a monetary award, but a two-week, all-expenses-paid trip to Mecca. It still didn't do the trick. Coelacanths, it appeared, could not be kept alive for any length of time at the surface.

In August 1975, a young revolutionary named Ali Soilih overturned the first Comoran government of Ahmed Abdallah, and set his country off on a roller-coaster ride of coups and counter-coups. Initially, Soilih brimmed with reforming zeal, but after a series of setbacks and determined opposition from the conservative elders, he descended into madness. He sacked the civil service and turned the running of the country over to teenagers with guns.

Two years later, a BBC cameraman, Peter Scoones, was sent to the Comoros to try to get live coelacanth footage for David Attenborough's *Life on Earth* series. After several frustrating weeks of trying to film in the ocean using an ROV (Remote Operated Vehicle) at a depth of one thousand feet, the closest he had got to a coelacanth was a stuffed model in the presidential palace. "We were coming back from a dive one evening, in the Comoran navy's only boat, and I looked at the island and it was red, as if it was sunset," he recalls. "Only we were on the wrong side of the island. I realized I was looking at rivers of molten lava—the volcano had erupted." Karthala, with its huge crater, dominates the island of Grande Comore. It erupts, on average, every twelve to fifteen years, destroying villages and depositing black lava into the ocean.

"Since I hadn't got any coelacanth footage, I thought I may as well film the volcano," Scoones continues. "I spent the night climbing through the rain forest, dodging clods of red-hot material which kept flying through the air. I got back to the hotel (appropriately, the Coelacanth Hotel in Moroni) at dawn, and no sooner had I got into my room, than there was a knock on the door." A coelacanth, he was told, had just been caught in a nearby village and was still alive.

"I rushed to the village to find it tethered under the shade of some pirogues. I carried it back into the water and tried to revive it by passing water over its gills. It was clearly exhausted, and one of its fins had been almost severed. It kept trying to bite my hand, which was actually quite useful, because then I could maneuver it into a good position to film. The only problem was that it seemed to want to stand on its head or swim upside down. Still, I got some good pictures."

Scoones's adventure was not over. A few days later, he was in Moroni, where he noticed young militiamen patrolling the streets. He walked past a two-story government building. Paper was floating from the windows to the ground, where he saw it being burned on a small bonfire. "I didn't really know what was going on," he says. "I asked some chap, who said it was a coup, but it didn't really look like much. Like everything else on the islands, it was pretty disorganized." Unbeknownst to Scoones, he had witnessed Ali Soilih's last attempt to brand his presidency—by burning the government archives. With the treaties, laws, and records went all the papers relating to coelacanth catches over the five years since the French had left. In coelacanth history, the Soilih years are a complete blank.

Ali Soilih was overthrown in 1978 by a mercenary force led by the infamous soldier of fortune Bob Denard, who had been hired by the first president of the Comoros, Ahmed Abdallah. For the next eleven years, Denard and his men were a constant presence on the islands, overseeing their security. In time, they came to be feared and hated by the majority of the population. Coelacanth specialist Robin Stobbs tells of going to the Comoros on a Smith Institute research expedition. "We were sitting in a fairly empty plane when a distinguished, silver-haired man came back from first class, asked who we were, and introduced himself as Colonel Bako [Denard's pseudonym in the Comoros]. It was only later that we realized who he really was."

The Smith Institute group was given an escort by the mercenary-led presidential guard. "I couldn't work out at first why the locals were so uncooperative," Robin Stobbs continues. "It was only when we met them without the mercenaries that they opened up."

Adventure and disorder were also on the menu for the Explorers Club, led by the indomitable coele-enthusiast, Jerome Hamlin. A Yale philosophy graduate and inventor of the first domestic robots, he had become intrigued by the coelacanth in 1984, when he joined a New York Aquarium beluga whale-hunting expedition as a photographer. "The procedure was quite fascinating, and at the conclusion I asked various aquarium experts and observers what the most challenging and scientifically valuable collecting project would be. They answered, 'The coelacanth.'"

It was as good as a lure for Hamlin. In 1986, he went to the Comoros for the first time, to sniff out the possibility of capturing a live coelacanth for the New York Aquarium. Two days after he arrived, he was having tea at the Coelacanth Hotel when a waiter announced that a coelacanth had been caught. "We were down the coast road in a blur, our hearts pounding," he recounts. "The fish was dead, but very fresh and its eyes still glowed. I have never seen such a beautiful creature. We bought it from the fisherman and persuaded the chef at the hotel to store it in his freezer. I was careful from then on not to order 'fish of the day'!"

A few days later, he was woken by a knock on his bungalow door and an excited whisper, *"Un coelacanthe vivant."* He was taken to the shore, where a fisherman was waiting with a living coelacanth tied to his boat. Hamlin climbed into the canoe and tried to keep the fish from bumping against the side. He then floated a folded rubbish bag over the fish's eyes to shield it from the rising sun. "As soon as I saw the coelacanth, all thoughts of photographing the creature flew from my mind: all I wanted to do was to keep it alive," Hamlin recalls. He summoned the French diver Jean-Louis Gerard, who swam down with the fish, and secured it to the ocean floor with a rope threaded through its lower jaw. Sadly, the coelacanth was unable to survive the adventure and died overnight. It was taken to the U.S. embassy freezer where it stayed, next to the body of a small American child who had died of malaria, until the next available flight, when it was transported to America in a coffin lined with pallets of dry ice flown over especially from France.

After two such near misses, Jerry Hamlin was hooked. He returned to the Comoros the following year with crates of equipment, including an enormous fish transporter tank, refrigerator units, oxygen tanks, portable generators, and a couple of float bags. At that stage, he still believed it was important to get a live fish in captivity, to learn how they lived and how to husband them. With a team from the New York Aquarium, he set out day and night to find one, with little success. "But there was no way I was going to give up," says Hamlin. "I was determined to keep coming back until I got a fish, and I started to put in place the structures for a more permanent coelacanth monitoring team. One afternoon in 1987 I noticed a large Comoran man—with the build of a silverback mountain gorilla—hanging about the hotel pool area. We got chatting, and it transpired that he worked there as a security guard several days a week and went by the name of Mombassa, after the town where he had been a boxer several years before. He had also acted in two John Wayne movies—he was the chief safari boy in *Hatari*." It transpired that Mombassa was a famous character on the Comoros, and had worked with a Japanese coelacanth-hunting expedition the previous year. Hamlin hired him as his local Mr. Fix-it, and Mombassa became his partner in adventure for the next ten years.

Hamlin returned to the Comoros in 1988 to try again. This time, he operated without the New York Aquarium. A series of misunderstandings had resulted in some major rows in the coelacanth world, which had prompted the aquarium to pull out of the project, leaving Hamlin more or less to his own (often extraordinary) devices.* He had become consumed with the idea that he had to save the coelacanth, and to that end, he changed his aim from capture to conservation. "The idea was to resubmerge the fish in steel cages to their capture depths, in the hope of resuscitating them," he explains. On five occasions, specimens were returned to depths of 30 to 50 meters (90 to 150 feet). One fish managed to survive for five days, during which time it was fed chicken by divers, before dying, possibly as a result of an eel attack. Another bit John-Louis Geraud on his hand; a third was lost after Mombassa fell asleep on the beach and failed to notice that its marker buoy had vanished. (This fish was rumored to have been stolen by the mercenaries.)

In 1989, Hamlin arrived in the Comoros the day after a coelacanth had been caught and resubmerged. He went straight from the airport to the village, where he found the fish already dead, with big yellow bruises on its lips, perhaps a result of banging its head against the cage. These obvious signs of stress prompted Hamlin to drop the cage idea and go into "passive mode." He installed a 700-gallon, cooled-down aquarium in a tent opposite Itsandra Beach, looked after by Mombassa. The first two fish that were intercepted upon capture were rushed to the tank, but both died before they arrived. A coelacanth caught in 1995 lived for ten hours in the facility, having survived for fifteen hours previously on the ocean surface. There were no further attempts, and in 1997 Mombassa was killed in a traffic accident, temporarily halting the possessed Hamlin's efforts.

Foreign expeditions continued, with the clear aim of trapping a live specimen. An American aquarium offered to pay $40,000 for a

---

*A colleague of Hamlin's had placed an advertisement in an angling magazine appealing for volunteers to monitor the coelacanth project in the Comoros. Various coelacanth scientists read this as an indication of the New York Aquarium's desire to fish actively for coelacanths. After an exchange of angry letters in the New York Times, the aquarium dropped out. Hamlin later made his peace with the scientists.

living coelacanth, while the Toba Aquarium of Japan launched the most ambitious—and certainly the most expensive—attempt yet to capture two live specimens in 1989. The multimillion-dollar project, sponsored by Mitsubishi, used a research vessel equipped with an ROV, specially designed traps, and fishermen from Japan and the Philippines, but they still failed in their objective.

It was around 1987 that a worrisome rumor started to circulate. It was said that the coelacanth had entered Chinese mythology, and venal Chinese herbalists were promising that a drop of notochord fluid would guarantee immortality. Coelacanths apparently were being picked up on the black market by Japanese middlemen and sold to Chinese practitioners who, the rumor went, were selling a single drop of the fluid for the incredible sum of $1,000. Since a coelacanth contains around three liters of clear, amber-colored oil, this would make them an extraordinarily valuable fish. Suddenly, it dawned on the scientific community that the coelacanth was in danger of extinction. Although no one could calculate the exact size of the coelacanth population, it appeared to be shrinking. What J.L.B. Smith had prophesied in his letter to *The Times* in 1956 was coming true: the "seemingly uncontrolled massacre" of coelacanths in the Comoros threatened to wipe out the species altogether. And their blood would be on the hands of the scientists. The emphasis, they realized, would have to change from collection to protection.

Sometime in the future, perhaps, the coelacanth could be studied in its natural environment—without endangering its continued existence—but it would take new men and new methods before this could become possible.

IX
DER QUASTENFLOSSER

In 1975, a young German scientist visited the Comoros to dive for the coelacanth. After several failed attempts, he turned to his wife and said, "Next time, I will come here with a submarine!"

He was not entirely joking. Since reading *Old Fourlegs* as a teenager in East Germany, Hans Fricke had been determined to see a live coelacanth. From an early age, he had been entranced by fish and marine life, and loved diving, which unfortunately he had little opportunity to pursue behind the Iron Curtain. "More than anything, I wanted to see a coral reef," he explains. "I knew I would never be able to while I lived in East Germany, so I had to escape." In 1960, a year after he had read Smith's book, Fricke—then a young navy cadet—left his family and boarded a train to West Berlin. The Berlin Wall was on its way up, and he knew the consequences would be serious if he was caught by the police who patrolled the train, checking everyone's papers. Searching for a hiding place, he ducked into a cabin reserved for women and children, where he saw a high-ranking army officer with his two young daughters. "I sat beside him and started talking. When the policemen came to our carriage, they recognized the officer's seniority and didn't stop. They must have thought I was his son."

Fricke made it to the west. He went back to college, as his eastern qualifications were not accepted in West Germany. At night, he sold newspapers in the city's nightclubs, bars, and brothels. "The women were all very charming," he recalls with a smile. "They were good to me because, like them, I had a street profession. I remember once, at about five in the morning, I was terribly tired but I still had fifty papers to sell. I went to a striptease place, and one of the dancers took pity on me and bought all my papers—without losing the beat."

With the money he earned from his night job, Hans Fricke went to the Red Sea in 1961 and saw his first coral reef. He was hooked. The next year, he returned—by bicycle across Europe to Greece, by ferry to Alexandria, and by bicycle again to the Red Sea. He cycled a total of ten thousand miles, picked up a debilitating stomach bug, and lost forty-four pounds in the process, but the struggle did little to diminish his wonder. From then on, he went back every year to study and photograph the exotic reef life.

He never forgot, however, his desire to see a living coelacanth. In 1968, he started working with the famous Konrad Lorenz at the Max Planck Institute for Animal Behavior in the small Bavarian hamlet of Seewiesen, and a year later, he went on a research trip to the island of Nosy Be in Madagascar. "There I met the French scientist Raphael Plante, and we dived for coelacanths along the continental edge of Madagascar. Of course we had no luck." They stopped off in the Comoros on their way back to Europe, and stayed in the Hotel Coelacanthe in Moroni, from which they dived again for the elusive fish—again with no success. It was six years later, on Fricke's next trip to the Comoros, this time with his wife, Simone, that he decided to return with a submarine.

The love affair between marine biologists and vehicles of the deep had really begun forty-five years earlier when William Beebe from the New York Zoological Society became the first person to conceive of a way to explore the ocean depths. After years of examining a small area of water near Bermuda, using nets to pull up more

than 115,000 animals—representing 220 species of marine life—
he had become increasingly frustrated at the limitations of his
method. The fish he sampled were at best the passing trade of his
ocean patch, and he knew his discoveries were little more than a
drop in the huge ocean. So he decided to invent a machine that
would take him into their world, to see firsthand where and how
these creatures lived.

In the early 1930s, Beebe designed the first bathysphere
(from the Greek *bathy*, meaning deep). It was in effect a steel ball,
specially created to withstand the great pressures of deep water. Its
walls were made of an inch and a half of solid steel, its windows of
solid quartz six inches wide and three inches thick, and it had a
hollow just large enough to accommodate two cramped people.
With its spidery legs and cable fitting, it stood as tall as a man,
with a small entrance hole sealed by a door that weighed four hundred pounds, and an outside spotlight that could be switched on
and off. Oxygen was stored in pressurized tanks, and carbon dioxide neutralized with the aid of an open tray of chemicals over
which Beebe periodically waved palm fronds in order to speed up
the reaction.

The sphere was lowered by cable into the often pitching
ocean while Beebe and a colleague sat inside, watching their world
turn an ever deeper shade of blue. From his first dive in June
1930, Beebe was captivated by a world that he was the first living
person to see. "On earth at night in moonlight I can always imagine the yellow of sunshine, the scarlet of blossoms," he recorded in
his colorful account of his underwater expeditions, *Half Mile
Down*. "But here, when the searchlight was off, yellow and orange
and red were unthinkable. The blue which filled all space admitted
no thought of other colors."

He saw all kinds of weird and wonderful creatures, many previously not known—or even thought of. There were fish and jellies
of extraordinary shapes, illuminated by their own flickering lights.
On one trip, he saw a huge fish with large eyes and a gaping jaw,
which it held open to reveal fangs glowing with what appeared to be
luminous mucus. Spots of pale blue lit its body, while two tentacles

drooped from its head, one tipped by a reddish light, the other blue.

In 1934, he descended to a record depth of half a mile—many times deeper than man had ventured before, either in a dome-capped suit or a submarine. There he saw a creature more than twenty feet long and very wide, seemingly without color, texture, or even a distinct outline, moving silently through the dark water. He was never able to identify the strange, serpent-like beast—to this day nobody knows what he saw—and he concluded that man's knowledge of the watery deeps was indeed still very limited.

Successive generations of marine biologists latched onto Beebe's invention and began to build more bathyspheres and submersibles, as they are now known, to explore ever deeper parts of the ocean, where they encountered a huge number of new and extraordinary species. This enthusiasm filtered into the public consciousness, thanks in part to Jules Verne's *Twenty Thousand Leagues Under the Sea*, which was inspired by the early bathysphere voyages, and to Jacques Cousteau, who became famous for his films of underwater life.

Hans Fricke started looking into the possibilities of hiring a submersible soon after his return from the Comoros in 1975. But it was three years later that he began to take his quest seriously. At a conference in Geneva, he met Jacques Piccard, who had descended to a record depth of 11 kilometers (6.7 miles) in his submersible. Piccard offered to take Fricke for a dive in Lake Geneva, and as they resurfaced, Fricke turned to him and said, "This is something I want to do in the future."

First, Fricke looked into the possibility of bringing Piccard and his submersible to the Red Sea, but it turned out to be far too expensive. "So I investigated whether I could purchase a submarine. But that was also too expensive," he recounted. "I soon realized that most of these machines were far too complicated, and needed specialists to service each of the parts. I wanted a machine which was simple enough to enable me to do my own maintenance. I started to contact people who built submersibles."

Among the strange and interesting people he met was a Czechoslovakian engineer based in Switzerland, Jaroslav Kahout.

Fricke went to his workshop near Zurich, and was impressed by his machining techniques and by his ability to improvise, "typical of a man from the East." Together, the two exiles from behind the Iron Curtain set about building a two-man submarine at the bargain basement cost of 130,000 DM (about £35,000), which the German magazine *Geo* had pledged to fund. "It was very, very basic," Fricke explains. "I used my imagination, and Jaroslav translated that into reality. We made dummies of the interior to find out the psychological limits to space—how small we could make it without going mad."

A yellow submarine, not unlike a beetle, was born and christened *Geo*, after the magazine. It had rounded, wing-like ballasts filled with pressurized air on top, and two thick, bug-eyed portholes, one at the front and one on top—like a skewed pair of spectacles for the very poor-sighted. There was a spotlight attached outside, a base for a camera, and a long, claw-like manipulator, which could be operated from inside and used to deposit and pick up materials from the ocean bed. There was just space in the cabin for two; the pilot sat behind the observer, whose eyes were on a level with the window. It was designed to be able to descend to a depth of 200 meters (660 feet). Hans Fricke was delighted with his new baby, which for the first few years spent most of its time in Europe and the Red Sea. "I hadn't forgotten about the coelacanth," he said, "but for that project we needed a mother ship, which wasn't easy to find."

One of his adventures to the Red Sea in the early 1980s involved building an underwater house. This Fricke lived in for eighteen days nonstop at a depth of 11 meters. It was a 26-metric-ton structure with a wet room, and a dry room into which air was pumped from shore. As well as observing new fish behavior and taking some wonderful photographs for *Geo* magazine, he experimented with putting free-floating objects into the water and watching as fishes began to congregate around them. "This was the first FAD—fish attracting device—which other people later commercialized, and which were used to great effect in the Comoros in the late eighties," he explains.

Fricke had built the house with a friend, Gerd Helmers, who did most of the steelwork. Fricke told him of his desire to take his submersible *Geo* to the Comoros to look for the coelacanth, and of the problems he had encountered in finding a suitable mother ship. Helmers, clearly a good friend, responded with the offer to build Fricke a yacht. The two-masted *Metoka* took longer to construct than *Geo* had, but was finally finished in 1985. With Jürgen Schauer, a colleague of Fricke's at the Max Planck Institute, on board, it sailed from its shipyard in London to Eilat in the Red Sea, where it met up with Fricke and *Geo* for trials. Jürgen then sailed on the three-month journey to the Comoros, where he joined Fricke and Raphael Plante on Christmas Eve, 1986, to look for coelacanths for the first time with a submersible.

"It was nerve-racking," Fricke conceded. "I had persuaded the German Research Council and *Geo* magazine to give us large grants for the expedition, but it was by no means a sure thing that we would succeed. It was an expedition into the blue. Of course, I had investigated the possibilities pretty carefully, and I was relatively certain that I would make it, but it was still a risk: if you fail with a research grant on such a scale, then you are lost for the rest of your life."

The signs were not good. The day after *Metoka* arrived in Grande Comore, the weather turned, and for the next weeks, the expedition had to battle tropical storms with their torrential rain and bucking seas. They were forced to spend days on not-so-dry land, sitting in a small café in Moroni, the crumbling capital of the Comoros, gazing at the angry, khaki ocean crashing against the harbor walls. Around them, life continued as normal: unfazed by the seasonal monsoon, Comoran women swathed in brightly patterned cloth sat in the old market in front of small piles of fruit and vegetables, with plastic bags providing an ineffectual shield from the deluge. Children skipped along the narrow alleyways of the medina, and old men in long white robes played fierce games of dominoes under a broad-leaved mango tree in the *place publique*. Eventually, the rain lightened enough for Fricke to dive for the first time in Moroni harbor, in front of the graceful twelfth-century Friday Mosque. Practically the whole city turned out to watch. "We

Hans Fricke (*right*) and Jürgen Schauer inside *Geo.*

went down to two hundred meters—*Geo*'s maximum depth—and we experienced what became a familiar story over the next three weeks: nothing—a sandy bottom, caves etched from volcanic rock, and very little marine life. It was not dark at that depth, just a little gloomy when the sun was not out."

They dived at numerous sites along the perimeter of the 57-kilometer-long island, spending long hours beetling around underwater. Still nothing. Occasionally, they saw coral heads, with delicate lattice-work fans fluttering with brightly colored fish and crustaceans. But most of the time they wended their way along dim and silent canyons, where few creatures chose to live. "The whole time we were expecting to see Old Fourlegs sitting on the sea bed," Fricke recalls. "At that stage, everyone thought they rested on their fins, like they were legs. I had a hunch that they didn't walk underwater, but I didn't think that all those famous professors could be wrong. It was just that their fins were not like those of a bottom-dwelling fish. I am not an expert in fish locomotion, but I have seen so many fish in my life, and it just struck me that it was rather fishy!"

At one point, they were diving off the ancient village of Iconi when Jürgen Schauer, *Geo*'s pilot, suggested that they should look into a cave. Fricke demurred, however: "A fish that big would not live in a cave," he said, so the caves remained unexplored. At each coastal village, they interviewed fishermen, and were impressed to discover that the fishermen's estimations of the ocean depth were extremely accurate. They also learned that almost all the known coelacanths had been fished up on the west coast of Grande Comore, and all had been caught at night. After examining the topography of the ocean bed on both sides of the island, Fricke came up with the reasonable hypothesis that the coelacanth favored the west coast because it was structurally more complex, composed of lava from relatively recent volcanic eruptions. He decided to concentrate their efforts on the western side.

"As the weeks passed, we became more desperate to see the fish. We were diving really intensively, even through a cyclone, which threatened to destroy everything. It was as if we were involved in a constant fight against the sea." Fricke's time ran out, and he was forced to return to Europe, having made forty dives without seeing a coelacanth. Jürgen Schauer was due to stay behind to continue diving for another five days. They had decided to dive only at night, since that was when the fish had been caught, and on the night that Fricke left—as he was sitting at the airport waiting for his delayed flight—Jürgen and a young student named Olaf dived off the coast near the village of Singani, site of Karthala's last eruption in 1977.

"I had a strong feeling about the area," Jürgen recalls. "It looked very mysterious—like the fish. We descended at 8:30 P.M., and at 9 P.M., we saw the first coelacanth. There it was, at the edge of the spotlight, *der Quastenflosser*. I held my breath. It was a very, very exciting moment, one of those special moments of a lifetime." He was captivated by the fish's large, shining eyes, its wide mouth, and its delicate dorsal fin, splayed like an antique paper fan. "It was now crucial to see how it would behave when faced by a large yellow submersible," he continued. "We approached it very slowly, then all of a sudden, right outside the window, it went into a headstand—just stayed suspended, nose down, waving its fins slowly and gracefully in the water, almost as if in slow motion. It was so beautiful, like a ballerina dancing."

Unfortunately, the video camera exploded in a puff of smoke when he tried to switch it on, so he took still photographs. After about twenty minutes, the current carried *Geo* too close to the dancing fish, cornering it against the rocks. With a quick movement, it squeezed past the submersible and vanished into blackness. As it knocked the metal, one of its scales came off, which Jürgen managed to retrieve with the manipulator. The submersible surfaced and he jumped into the water and slipped the scale into his pocket before hauling himself onto *Metoka*. "The mood was very negative on the ship. The captain and crew had almost given up hope and were just going through the motions. Olaf and I decided to keep quiet about it for a bit, so we just got on with things and pretended nothing special had happened. After about half an hour I grabbed the skipper and said, 'We got it!' He didn't believe me, and of course, we had no film footage to prove it, so I took the scale out and showed it to him. I still have that scale in my treasure box."

The next day, Jürgen managed to contact Hans Fricke in Germany, who, not surprisingly, was delighted to hear the news. "Actually, I was more relieved than anything. The operation was saved and now we would get the money for a second expedition. It didn't matter that I hadn't seen it myself: I was just really happy that we had found one." Over the succeeding nights, Jürgen saw two more coelacanths, which he managed to film with a borrowed camera. "When I saw Jürgen's footage for the first time," Fricke relates, "a big stone fell out of my heart. It was great, really fantastic.

"I suppose, if I was to be honest, I would have to say that the search for the coelacanth was foremost an adventure and a challenge. I knew they must be down there somewhere, and it was fun and exciting to look for them. To find them and succeed in something scientists have been trying to do for half a century was a thrill, a real thrill."

The team returned in April 1987, and again encountered foul weather. They dived only at night, when the conditions were at

"Nico," the first photograph of a live coelacanth.

their worst: huge swells and high waves crashed against the side of the ship, making lowering *Geo* into and out of the water difficult and dangerous. They spent weeks moving slowly through the black water, between steep and barren lava cliffs. There Fricke saw his first coelacanth, a small, near-perfect fish that he christened Nico, after his son. Over the next weeks, they watched and followed three more fish—shadowing one for six hours.

Fricke was fascinated to study the coelacanth's movement. At first glimpse, it looks as if it is waving its many fins in an uncoordinated fashion. But as he looked closely, he realized that the fish moves its fins in a diagonal fashion—left front fin with back right—a gait very similar to a horse trotting, or a lizard walking. He also exploded one of the most enduring coelacanth myths: that it uses its lobed fins as feet to walk on the seabed. "Though we observed several individuals resting with their fins braced against the sea bottom," he wrote in the *National Geographic* magazine after returning from his 1987 trip, "we never saw any of them walk, and it appears the fish is unable to do so." Old Fourlegs, it seems, was not so aptly named.

They had gathered enough footage to make a magical film, which showed the coelacanth, at last, in its natural habitat, swimming with a grace unexpected in a fish so large and—outwardly at least—so cumbersome. It looks and moves unlike any other fish, sculling its splayed fins like a Japanese fan dancer, standing on its head like a gymnast. "I always say it is a creature that doesn't belong in our marine world," said Fricke. "It is a very special fish."

His film and observations, however, only increased the coelacanth's mystique, raising as many questions as it answered. Where did it spend the day? Why did it stand on its head? How did it feed and reproduce? How many coelacanths were there in the Comoros, and could they live anywhere else? Fricke knew he would have to return if he was going to try to answer these questions. He suspected that the fish spent their days in deeper water, and if this was the case, he would not be able to see them in *Geo*. He determined to build another submersible, capable of descending to greater depths. As soon as he got back to Germany, he started to draw up designs, but it would be two long years before he could return to the Comoros to resume his quest.

In 1988, Fricke took up an invitation from Margaret Smith to deliver a lecture on the coelacanth in Grahamstown. Since conceiving his expedition to dive for coelacanths, he had been in close contact with Margaret Smith and her team at the Institute of Ichthyology in Grahamstown, still a scientific hub of the coelacanth world. Having read *Old Fourlegs* as a boy, he was keen to meet J.L.B.'s widow in person for the first time. In the two decades since her husband's death, Margaret Smith had become famous in her own right.

"I have always been one to count my blessings," she once wrote. "I realised what a tremendous privilege it was to share the life and work of one of South Africa's great men—realised it every day of the twenty-nine and three-quarter years of our marriage without any regrets. I knew that with the great difference in our ages (19 years) I would probably be left a widow—when I married him he was given five years to live—we have had nearly thirty, and few couples can have had a more interesting and productive life than ours."

She seamlessly assumed her late husband's mantle and responsibilities. People who knew her both before and after his death talk of a visible "blossoming." She cut off her severe bun, and moved from their small, rather awkward house into a lovely old colonial manor, which she shared with her sister, Flora Sholto-Douglas (known as Aunt Flo, who had been heartily disapproved of by J.L.B.). Her first goal was to consolidate the J.L.B. Smith Institute of Ichthyology. As she wrote to J.L.B.'s estranged sister, Gladys, soon after his death: "Len refused to have a new building put up for the Institute while he was still alive. He preferred to spend the time on his research work. . . . And now he has not left me—I am continuing with his work as we had always planned I should. For 22 years I worked day and night at his side, and I hope to finish what we started before I die."

She visited similar institutes around the world, and returned to Grahamstown to work closely with the architect to make sure the establishment that bore her husband's name would be perfect. She became its first director, and her energy and perseverance, together with the Smith name and the coelacanth's fame, ensured she was able to raise the requisite funding. In later years, when the institute was thriving, she campaigned for it to be credited as a national museum, with all the financial benefits that entailed. At its head, she quickly became a figure of great stature in her own right: a world-class ichthyologist, an extremely talented illustrator of fishes—over two thousand accurate and intricate drawings and paintings—and an international ambassador of science.

Margaret Smith was a larger-than-life figure—both in her professional and private life. She took up singing and music again, and she and Aunt Flo held hysterical joke parties, where the humor at times became embarrassingly ribald, reported Mike Bruton, who succeeded Margaret Smith as director of the J.L.B. Smith Institute: "It was almost as though she needed to chip away at the academic facade."

At work, she was adored by her "children," and her office was perennially crowded with people, as well as donkey harnesses and boxes of stamps for the Red Cross, something J.L.B., with his exaggerated work ethic, would never have tolerated. She took to committee life like one of her fish to water, and was often to be seen scurrying between meetings, clutching her knitting. "She would readily end an important meeting to cuddle the baby of a visitor," Bruton recalls, "but she was a formidable foe to any government official or academic who criticized her management style. To her, happiness and harmony were of paramount importance. Tea was a loud and joyous occasion as she recounted the hilarious happenings on her early expeditions."

"Life was always exciting when one was around Margaret Smith," her secretary, Jean Pote, remembers. "She was always fired up and enthusiastic about everything, and it seemed to rub off on all those around her. The small things in life were very important to her: if, for instance, she heard of a sheepdog show in the countryside somewhere, she would sweep off, taking along a handful of her staff for the outing. She gave a huge amount to charity, mostly on the quiet. Beggars would come off the street and she would interrupt a meeting to lend them money, which she would write down in a little black book—always failing, however, to chase it up if they didn't repay her. She made opportunities for people. Especially for underdogs."

Her generosity was repaid in kind. When her well-known gray Ford was stolen and taken to the nearby township, the thief immediately returned it when he learned it was hers. And when she got a speeding ticket in Somerset Street, right outside the institute, the cop tore it up when he saw who the driver was. According to her son William, "Mom was a saint, the only person I know who could get into a lift full of strangers on the ground floor, and by the third floor, she would know their names and family history and they would all be friends. It was like she developed to be a foil for Dad—and she paid a price for it. When she was growing up and at university, she was tremendously career-minded, but when Dad was alive she had to play second fiddle to his greatness. You can't change nappies and be a great scientist: she was very good at nappies. But she was also great, and she got her greatness through people."

Margaret Smith painting fish for *Sea Fishes of Southern Africa*.

Margaret Smith embraced life with the same boundless energy as J.L.B., but without the accompanying grumbles. She continued to go on fish-collecting expeditions for as long as her health permitted. According to Robin Stobbs, who accompanied her on her last trip to Mozambique, she never flagged: "The endless long days of collecting under the tropical sun, sorting, labeling, pinning out fishes, photographing some and drawing others, preparing for the next day, often only crawling to bed well after midnight, was a routine that Margaret followed with the unflagging enthusiasm which might have daunted someone half her age." At the age of sixty, she learned to scuba-dive in Hawaii.

She became a full-fledged professor in 1980, and two years

later, retired as director of the Smith Institute to concentrate on her last great project: *Smith's Sea Fishes*. This was to be a monumental and definitive tome, successor to the earlier *Sea Fishes* books, which she coedited with Phil Heemstra, incorporating the work of seventy-two scientists from fifteen countries.

By the time the project began, her health, once so reliable, had started to fail. A bad attack of arthritis resulted in an operation to have one of her knee joints replaced. In 1985, she contracted pneumonia, septicemia, and bacterial meningitis. She was in a coma for several hours and not expected to live. However, as Phil Heemstra recalls, "Her indomitable will and determination to finish the book pulled her through. She just refused to die. And even after she was confined to a wheelchair, she continued to be an active presence, driving herself around in an automatic car, lifting herself painfully out of the chair."

Shortly after the book was finished, Margaret Smith contracted leukemia. She was in the hospital, and in extreme pain, when Hans Fricke arrived in Grahamstown to deliver his coelacanth lecture. It was a great success; he showed his new footage, and was amazed and gratified that some of the audience cried when they saw the film. The next day, Fricke went to see Margaret Smith in the Port Elizabeth Provincial Hospital, and to introduce her to the living, swimming coelacanth. "It was the only time I met her," Fricke recalls, "but she made a very strong impression on me. I thought she was a remarkable person. With her personality and her knowledge, to have lived with such an egocentric husband must have been terrible."

He projected the first graceful images of the swimming coelacanth onto the bare white wall facing her bed. She was enthralled, and by the time it was finished, she was in tears. "She said that seeing the live fish had completed the circle of her life, and she was now ready to die," Fricke remembers. "She had been very depressed about her illness, but seeing the coelacanth made her incredibly animated and revitalized, emotional and fired up about it. She said she would take the memory to J.L.B."

Hans Fricke returned to his submersible blueprints, and six weeks later, Margaret Smith died.

X

JAGO

Hans Fricke had decided that, with all they had learned from *Geo*, he and Jürgen could build the new submersible on their own. To be able to descend to greater depths, the new machine would have to be more sophisticated: stronger and safer, and inevitably, more expensive. By doing it themselves they would save a fortune.

Their center of operations was a small wooden shed at the edge of the Max Planck Institute car park. As with *Geo*, a dummy was built, and gradually the body of the machine was created around it. "If you do everything on your own, with a small group of people, you rapidly become an expert," Fricke explained. "This is what happened to us." Quite by chance, he met a submersible builder on the road outside his house one day, who agreed to help with the technical specifications. Soon the new improved yellow submarine was ready.

Jago was named after a deep-sea shark (known in English as Iago) with the same color eyes as the coelacanth, which lives at 400 meters—the maximum depth *Jago* was designed to withstand. It was slightly larger than *Geo*—a meter and a half wide, and two and a half meters long—with bigger windows. Like *Geo*, it was free-swimming, fitted with powerful spotlights, an efficient manipulator, and a radio communication system with the surface. Fricke took his new submersible for a test dive in Lake Geneva with Jacques Piccard and his submersible, *Forel*, on standby. *Jago* passed its tests with flying colors. Fricke was delighted, and took the team to the Red Sea, and from there, back to the Comoros. They arrived towards the end of 1989 with the express intention of finding—and filming—a coelacanth by day.

It did not happen immediately. "In our imagination, we were sure we would find coelacanths straightaway below *Geo*'s 200-meter limit," says Karen Hissmann, who joined Fricke's team in 1988. From scientific analysis of the coelacanth's blood, it appeared that its best living conditions would be at temperatures of 15 to 18 degrees centigrade, occurring below 200 meters in the Comoros, and catch records seemed to tally. "Our first dives down to 400 meters opened up a new world to us: giant white, funnel-like or branched glass sponges, bizarre, beautiful, fascinating formations—but no coelacanths."

They noticed how the underwater landscape changed as they descended. Below 200 meters, the dramatic canyons gave way to shallower slopes of ash-colored sand. They encountered new and interesting fish, and at 400 meters, they found themselves in a dusky, barren landscape. They calculated that at that depth, they were resisting pressure of 3,600 metric tons: "This incredible thought reminded us of the extreme living conditions surrounding us," Karen Hissmann continues. "After a long search we were forced to admit that this wasn't coelacanth country."

They decided to go back to the area where they had seen coelacanths two years earlier, and on November 5 at 9:45 A.M., their efforts were rewarded when they caught sight of a coelacanth standing on its head, silhouetted against the entrance of a cave. For the first time, Fricke was able to see the graceful fish in daylight. "As we approached, it retreated into the cave," he recounts. "We peered inside, and to our great excitement, there were three more pairs of large eyes, glowing in the half-light. We saw that the coelacanths were resting quietly together, close but not touching. At that moment, we realized where they spent their days, and why we hadn't seen them before. And the irony was that the cave was at a depth of 196 meters—within the diving limits of *Geo*."

Now that they knew what to look for, they quickly found many more coelacanths, sheltering in caves, protected from predators and strong daytime currents. There were never the same number in each cave: sometimes the fish were in groups of up to ten; at other times there would be a solitary individual. They saw three old friends from the 1987 expedition, which they recognized from their distinct white markings, and noted one in six different locations. They spent hours upon end in their yellow steel capsule, just watching. They soon began to get a picture of the coelacanths' daily routine: resting hidden by day, then, shortly before dusk, leaving the safety of their communal caves to forage singly for prey. The coelacanth, it appeared, hunts alone.

In order to get a better idea of the fishes' movements, a small "pinger," a radio transmitter that could be tracked from the submersible or the surface, was shot into the side of each of eleven coelacanths. Fricke and his team took turns shadowing the large creatures as they drifted along, riding the currents like an albatross in the air. The coelacanths appeared to be equally happy swimming forwards or backwards, standing on their heads, or lying on their backs—their fat-filled swim bladders enabling them to maintain neutral buoyancy in any position, hovering in the water easily, weightlessly, with only the slightest gentle movement of their fins.

The headstands puzzled Fricke. Few fish are known to perform them, certainly not to maintain the position for several minutes without apparent effort, "as if they were all auditioning for a job in the circus!" Each time the submersible approached, the coelacanth would slowly lift itself into the vertical position, gently waving the tiny central lobe of its tail from side to side. Fricke became convinced that this behavior was somehow connected with the coelacanth's ability to locate prey by detecting changes in the surrounding electric field—using the rostral organ, the jelly-filled cavities in the snout believed to act as an electro-detection system. He decided to experiment by emitting weak electric fields—mimicking an approaching fish—in the water. Sure enough, as soon as

A coelacanth stands on its head.

the coelacanth picked up the currents, it swiveled onto its head. While the results were not conclusive, they were plausible. In 1996, Fricke identified electricity-generating organs in the tail fin that appear to add weight to his theory: the headstand position, perhaps, enables the coelacanth to detect prey with greater ease.

They photographed every coelacanth they saw from both sides, christened them (Nico was the first to be named), and produced an Identikit for each individual, clearly showing its unique pattern of white splodges on the dark scales. At last count, they had records of 108 different coelacanths. "Each time we go back, we dread not finding one of our particular friends," says Karen. "When we see a dead fish we hope we will not recognize it."

They conducted a thorough survey of the coastal region in an attempt to estimate the coelacanth population size, and to map their caves. It was not a quick or an easy task. Coelacanth caves tend to have relatively small entrance passages, which arch into larger internal wombs. Only by lying on the floor of the submersible and looking through the upper edge of the curved front window could they watch and film the beasts at home. Even if the entrances were larger, the coelacanths would be hard to see: the spots of white on steel-gray scales melt into the mollusk-studded lava caves, providing near-perfect camouflage. "It is very, very difficult to see a coelacanth underwater," Fricke explains. "If you are not acquainted with coelacanth caves you will never see one. And you need an extraordinarily small submersible."

On an earlier visit in 1987, Fricke and Raphael Plante had encountered an expedition team from the Smith Institute, led by Mike Bruton. They met up one evening, along with Eugene Balon, in a grimy, flea-bitten restaurant in Moroni. By that time, over 140 coelacanths were known to have been caught since 1952, and by their estimates, there could be as few as several hundred remaining. As they discussed their separate findings, they realized that there was going to have to be an international effort if the coelacanth was to be "saved." They set up the Coelacanth Conservation Council (CCC), with the aim of gathering as much information as they could, and making it available to as wide an audience as possible. This, they hoped, would stimulate interest in the coelacanth, leading to a release of funds.

The principals of the CCC embarked upon a worldwide lecture

tour to raise awareness for the plight of the coelacanth. "All the time, we bashed heads against the furry mammal brigade," Mike Bruton complains. "It is much easier to raise money for the soft and fluffies than for the wet and slimies, especially if, like the coelacanth, they live so deep and can't be seen in their natural environment, or even in an aquarium."

They didn't have much success in raising money, but in 1989, after strong lobbying by Fricke, the Convention on International Trade in Endangered Species (CITES) slapped its strongest protection order on the coelacanth. It was listed under Appendix One (along with blue whales, snow leopards, Sumatran tigers, and six other species of fish, including *Neoceratodus fosterii*, the Australian lungfish), which declares it to be a severely endangered species and strictly forbids any trade in coelacanths. This was considered to be a tremendous achievement—if it could be enforced.

President Abdallah, however, for reasons of his own, refused to sign the CITES accord, leaving the door open for the Comoros to export coelacanths, dead and alive. Only a few months after the coelacanth was elevated to Appendix One, the *Jago* team ran into a head-to-head confrontation with the multimillion-dollar expedition from the Japanese Toba Aquarium, intent on trapping a live coelacanth. Fricke has always been vehemently opposed to the concept of having a live coelacanth on public display—even if it were proven to be easy and safe to catch them. Before embarking on the 1989 expedition, he had protested against Toba's project, and succeeded in provoking President Abdallah to issue a decree prohibiting the export of live coelacanths. But without his formal signature on the CITES accord, the Japanese paid little heed. They continued to lower their steel mesh catch cages.

Eventually, Fricke snapped. After a series of underwater encounters with the vicious-looking cages, he decided to take action. He prepared two laminated cards, one with a picture of the coelacanth above the legend: "Coelacanths—Let them where they are!" The second identified *Jago* as the sender. One night, they took *Jago* to a cage, and with the manipulator, carefully hooked the cards into the mesh. Shortly afterwards, the unsuccessful Toba team was ordered home— apparently at the express wish of the Japanese emperor.

"We need a live coelacanth in captivity," said Mike Bruton from his base at the Two Oceans Aquarium in Cape Town. "There are some vital gaps in our knowledge of population dynamics; their age and size at maturity, number of young produced, gestation period, growth rate, final age—some of which could only be obtained from observation of captive specimens." He believed a captive coelacanth would raise public awareness of a rare and beautiful beast, which would lead to an injection of funding for research, and conservation activities in the Comoros. "Once we know how to catch them, then we could increase the number in captivity, and perhaps breed from them. I am sure that we could easily get one, with the help of Fricke and his submersible. But he refuses to cooperate."

Hans Fricke has steadfastly maintained his opposition. "It is a mad idea!" he exclaimed. "Nobody has seen a juvenile. Nobody knows about the sexes. Nobody knows how they do love, and under what circumstances. And they think about putting one in a aquarium to breed? It is ridiculous! A South African aquarium already claims to have the historical right to be the first to put a coelacanth on display—but this is nonsense: nobody has the historical right to a fish! And if one aquarium gets its coelacanth—then what? Everybody else wants one, creating a run on the small coelacanth population. No, I will not help!"

It is the small—and, he believes, diminishing—size of the coelacanth population that worries Fricke most. The more he finds out about the coelacanth and its habitat, the more he despairs of its continued survival. Almost constantly, the fish is being subjected not only to harassment by scientists from around the world, keen for their own specimen, but also to increased fishing activity by the growing Comoran population. Each time they have been to the islands, Fricke and his team have conducted a rough census. Between 1989 and 1991, the number of coelacanths they counted in the census area appeared to remain fairly stable, indicating a total population around Grande Comore of somewhere below 650. By 1994, however, the number seemed to have declined drastically—a result, Fricke believes, of increased fishing pressure by the economically stressed fishermen.

Some coelacanth scientists, however, regard Fricke's population numbers as unduly pessimistic. Robin Stobbs believes the population could be many times greater than Fricke's estimates. "He has concentrated his calculations on the areas where the fish are caught, without really looking elsewhere," Stobbs maintains. "But if you examine it in terms of fishing methods, you see that, for instance on the east coast of Grande Comore and on Mohéli, Mayotte, and Madagascar, they don't fish in the same way, so they are never going to catch a coelacanth. And although there has been an increase in the total number of fishing *galawa* in recent years, there has actually been a reduction in night fishing activities, which is good news for *Latimeria*. I believe there could be many more fish out there, in no danger of being caught at present. After all, we don't yet know the ancestral home of the coelacanth: where was it from the time the fossil records disappeared 70 million years ago until the Comoros were created 60 million years later? Surely not paddling about over the deep ocean waiting for a volcanic island to pop up from below!"

"If you examine the socioeconomic situation in the Comoros," explains Fricke, "you see why the Comorans have to eat every little creature from the sea. The reef is virtually fished out, and they are having to go after deeper water resources. While they do that, they catch the coelacanth by chance. Who can blame the poor devils? Their survival is more important than the survival of a fish. People have to eat. It comes from our aggressive way of handling our planet and really it is very sad, but this is the way evolution works, and unfortunately at this time we are a constraint which is terrible."

In 1994, five years after the CITES accord was promulgated, the Comoros signed the accord, whereupon the official market in coelacanths collapsed. It is suspected, however, that there is still a black market, at least when the right buyers are in town. Some reports estimate that coelacanths can still fetch up to $2,000—five times the annual average salary of a fisherman, and very hard to resist. But, as many Comoran fisherman testify, catching a coelacanth these days is no longer an automatic lottery win.

Ahmed Bourhane, a fine-featured, shaven-headed fisherman from the village of Mindrahou, remembers only too well the time he caught his *gombessa*. "It was 1995," he recounts. "At about 11 P.M., I felt something take the bait, but at first I didn't realize what it was. When it came up, it was moving like a woman in bed. It was very heavy, and I only managed to put a big hook through its mouth on my third attempt. I touched its skin and realized it was a strange fish, but it was dark and I didn't know it was *gombessa*. I called to the other fishermen who had lights, and when they came over, I realized what it was.

"Oh, I was very happy because I believed I was going to be able to do my *grand mariage* and go to Mecca. Before, when fishermen from my village caught a *gombessa*, they sold it for a lot of money. I put it in my boat, and did everything I could to keep it alive. It didn't struggle." It was Sunday and there were no taxis, so he hired a car and took his fish to the museum in the capital, Moroni. They said they didn't want it. He took it up to the Galawa Beach Hotel on the northern tip of the island, but they didn't want it either. Nor did the Chinese embassy. "I had the fish in the car the whole day and spent 40,000 CFA (about $75), but no one would buy it. So I drained the oil, which I give to people when they are sick, and grilled the fish. It didn't taste too bad."

News travels fast in the Comoros—and the faster and farther it travels, the better the coelacanth's chances of survival. Even Hans Fricke is encouraged by an increasingly organized grassroots conservation movement in the Comoros, which employs somewhat unorthodox methods to get its message across. The Association pour la Protection du Gombessa was formed in 1994 and formally inaugurated four years later, on February 28, 1998, at a colorful ceremony in a whitewashed classroom, seemingly stranded in the middle of a lava-rock plain. Fishermen and elders from twelve villages, dressed in their best clothes, sat straight-backed behind old wooden desks; outside, women draped in brightly patterned cloth danced in circles, ululated, and sang about the coelacanth. "*Gombessa* is a wonderful fish," they sang. "The world likes *gombessa*—it is better to protect it and not to fish it, in order for our children to know it."

The proud president of the association, a veteran fisherman named Hassane Djambaé, opened the proceedings. Beneath his crimson fez, his face was a study of commitment. "We are here today to talk about how to protect *gombessa*," he proclaimed. "It will not be easy. Fishing is our culture, and this is one time when we have to teach fishermen not to fish." He was followed by a young boy who sang verses from the Koran.

"The aim of the association is to protect the coelacanth and also the bottom of the ocean," explained Said Ahamada, a young Comoran ecologist and member of Ulanga, the national environmental conservation group that has been a strong impetus behind the formation of the Coelacanth Protection Association. "Comoran fishermen do not make much money—on average about 2,000 CFA ($4.20) a day. They need help. It is hard for them to understand the global significance of the coelacanth, and so we stress the socioeconomic benefits they might accrue from its conservation."

The association's ultimate goal is to get the waters around Itsoundzou—where the highest concentration of coelacanths are thought to live—declared a national marine park, with enforced prohibition of fishing. Hans Fricke wants to plant a permanent camera at the mouth of a coelacanth cave, to transmit live footage to a coelacanth information center on the shore. "It must be something exciting, something extraordinary," he stresses. "You can't touch the animal; you can't dive to it; you can't have a tourist submarine because it would be too expensive, and wouldn't be able to get close enough. So the only way to see a live coelacanth is through real-time video images." He is convinced that the Comoran tourist trade would make the center economically viable. All that is needed is the money to fund it.

The Coelacanth Protection Association is pushing this plan, telling the fishermen that if they leave the coelacanth alone now, they will benefit in the future. Some enthusiasts have promised dire consequences to those who bring back a coelacanth; the news is spreading that far from receiving fame and riches from the *gombessa*, the fisherman who catches one will instead be cursed by a witch doctor. Already, several fishermen have reportedly cut loose a coelacanth rather than bring it to shore. Fricke thinks this the best news he has heard—"But, but, but . . . Can the coelacanth survive being caught and released? They are pulled up through the warm water, totally distressed and exhausted, there is a buildup of lactic acid, damage to the inner ear, and the respiratory system is severely stressed. On the several occasions when newly caught coelacanths have been released back at depth, they have not managed to survive. It has nothing to do with pressure, as the fat-filled swim bladder ensures that it is not susceptible to changes in the water pressure—but still, I fear the stress of fighting a hook leaves the fish with little chance. But maybe they are tougher than we think? Maybe if they are brought to the surface and immediately released, they will survive?"

Jerry Hamlin of the Explorers Club is convinced of it, and has staked his own money and energy as proof. After the infamous Mombassa was killed and Jerry discovered that his resuscitation tank had been dismantled, he decided to try a different tack in his ongoing battle to save the coelacanth. From the headquarters of the Coelacanth Rescue Mission, a picturesque former inn in wooded Greenwich, Connecticut, which he shares with an albino Burmese python, a Colombian red-tailed boa, an Indian Star tortoise, a Parson's chameleon, and a bald cockatoo, he set up a prizewinning coelacanth website. "I ran a 'Save the Coelacanth' contest on the dinofish.com website, offering a $500 prize," he recounts. "For months I got the expected suggestions of cloning, fish farms, artificial reefs, and so on—all hopelessly expensive and impractical. Then one day, I got an e-mail from a Dr. Raymond Waldner, a biology professor in Florida. He mentioned a technique that was used to release deep reef fish without causing them the fatal stress of having to swim back down. A barbless hook with a weight attached to the shank is inserted upside down in the fish's lip. A line is attached to the curve of the hook and the fish lowered by that line to the bottom. When the line is jerked up, the barbless hook pulls out of the fish's mouth. The fish is free on the bottom and the rig is pulled back up to the surface, where it can be used again."

Hamlin read the e-mail with interest, and tried to work out how it could be applied in the Comoros. "Then the coin dropped. I devised a way to make the rig extremely light and small, so light and small that it could be placed in a pouch and sewn to a T-shirt. The T-shirt could then be distributed to fishermen, and instructions for the system could be printed iconically on the back of the shirt. What an advantage: because this way the coelacanth's time in the warm water of the ocean surface could be reduced to a few minutes. Dr. Waldner got the $500. Then I set about designing the shirt, manufacturing the rigs, and sewing on the pouches. We produced about seventy, and the first batch was air-mailed to the Comoros, at a cost of over $1,000. The T-shirts were distributed to the fishermen of Itsoundzou in August 1998, by Said Ahamada, while hundreds of Internet contributors bought kitless T-shirts over the Web, helping to fund the project."

Eighteen months later, there is no confirmation of the method's success, but Hamlin is still hopeful. If it is proven that coelacanths can be painlessly resubmerged, he plans to send small fish tags to the islands, which could be used to mark released fish, so that they would be noticed on recapture, or observed from a submersible, indicating that the fish did recover. "I just hope this works and the coelacanth is saved, so that I can get on with my life," he says.

In 1991, an event had occurred that gave hope to some of those who feared the coelacanth population was in danger of extinction. Sometime in August, a huge female coelacanth was caught in the trawl net of a Japanese ship off the coast of Mozambique. The 1.79-meter, 98-kilogram fish was frozen on board, and in December 1991, handed over to the Natural History Museum in the capital, Maputo.

True to form, on Christmas Eve, the J.L.B. Smith Institute in Grahamstown received a fax informing them of the coelacanth's capture, and a few weeks later, Mike Bruton (at that time still director of the institute) and Hans Fricke went to Mozambique to investigate. Bruton was met at the airport by the museum's direc-

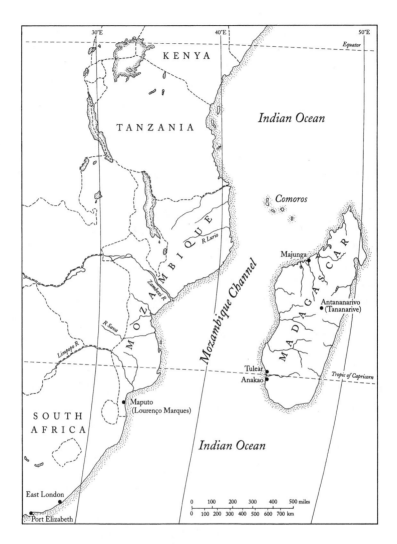

Islands of the southwestern Indian Ocean.

tor, Dr. Augusto Cabral, an extraordinary character who had single-handedly kept the museum going throughout his country's long civil war. Cabral reassured him that it was definitely a coelacanth—only the second to have been trawled, and indeed to have

been found anywhere but the Comoros. The bad news was that—
in another uncanny repeat of the events of 1938—lacking the facilities to keep it frozen, Dr. Cabral had been forced to dissect the
fish and discard its internal organs. The good news, however, was
that he had found twenty-six perfectly formed coelacanth pups
inside, which he had managed to preserve.

The Mozambique fish threw a slew of theories and assumptions into the air. Not only did the place and method of its capture
revive former conjectures that the coelacanth inhabited a much
wider geographical area—and that the East London specimen was
not necessarily a stray—but it also threw oil into the simmering
debate about the coelacanth's reproductive processes.

Until that point, all estimates of coelacanth population
dynamics had been based on the American Museum's specimen,
with its five embryos. Then along came the Mozambique fish,
quintupling the potential birth rate: if coelacanths could produce
up to twenty-six young, then perhaps they weren't as endangered
as we thought. It also destroyed, once and for all, Eugene Balon's
theory of uterine cannibalism.

To the skeptics and pessimists of the coelacanth world, the
Mozambique specimen wasn't such a big deal. Any live-bearer—
even one that produces twenty-six offspring—is a slow reproducer,
and it was entirely possible, they argued, that like the East London
fish, this coelacanth had also got caught in the southerly current
and drifted from the Comoros to the waters off Pebane where it
was caught.

There had long been rumors that coelacanths were to be found off
Madagascar. As far back as 1982, the government issued a miniature sheet of stamps depicting the coelacanth in its natural habitat.
Four years later, a specimen purported to have been caught locally
was displayed in Tamatave (now Toamasina), although it later was
believed to have been bought in the Comoros. Jerry Hamlin spent
several weeks traveling the length of the northeastern coast, inter-
viewing fishermen, but was unable to come up with concrete proof

The Mozambique coelacanth pups.

of a Madagascan coelacanth. The government, however, seemed fairly sure, and in 1993, issued a new coelacanth stamp—this time with a background of rocky overhangs and caves.

Two years later, their faith was justified. On August 5, 1995, a 32-kilogram coelacanth was hauled up in a deep-set shark net off the village of Anakaó, 33 kilometers south of Tulear on the southwest coast of Madagascar and 1,300 kilometers south of Grande Comore. It was caught by three young fishermen, none of whom had ever seen a coelacanth before. When they pulled up the net and saw their unusual catch, they were terrified and nearly threw it straight back in the water. The Malagasy follow a complicated system of taboos, or *fady*, which relate to their belief in mysticism and the spirit world, and which are rooted in their Indonesian ancestry. Anything that is strange or unknown is automatically *fady*, and it is prohibited to touch or handle it. To the Malagasy fishermen, the coelacanth, with its strange "limbs," must have seemed a freak of nature, and as such to be treated with extreme caution. Breaking a *fady* is a potentially fatal business. Nevertheless, the owner of the boat, ZeZe, decided to risk it and take the coelacanth back to the village to show to the mayor, Mr. Regis Robinson, an experienced fisherman.

Regis Robinson was just as confused, however, and the fishermen were on the point of cutting up the coelacanth to use as bait when a passing Frenchman recognized it as *Notre Coelacanthe* and bought it for 20,000 Malagasy francs ($6). He took it in his boat to the Museum of the Institute of Fisheries and Marine Sciences in Tulear, where it was preserved and exhibited.

Perhaps inevitably, this was not enough for everyone. Some skeptics suggested that it too was a stray, carried on the currents from the Comoros. But when in 1997 another specimen was caught, apparently by the same fishermen, in the same location and under the same circumstances, the coincidence was too great.

"It could very well be that Madagascar is the ancestral home of the coelacanth," says Robin Stobbs. The huge island—third largest in the world—was once part of the supercontinent of Gondwanaland, before splitting from the African mainland around 60 to 70 million years ago—at about the same time as fossil evidence of the coelacanth disappeared, making it far more ancient than the relatively new Comoros. "It was only very recently that fishing techniques changed in that region of Madagascar, and instead of using deep long lines to catch sharks, the local fishermen started using nets, into which this unfortunate coelacanth had blundered," Stobbs continues. "Until recently, few fishermen risked dangerous night fishing, far from shore, where they might drift out to sea and die. The bodies of the dead are treated with extreme veneration on Madagascar, and the supreme *fady* is for your body to go unburied—this reflects badly on your family and entire community. So perhaps the coelacanth has been there all along, but it has taken a change in fishing techniques to catch one."

Perhaps the coelacanth has been living quietly even farther away, all along? Over the years, a series of clues has emerged that seem to indicate that the fish's home range is wider than believed. Some of these clues were undoubtedly red herrings, and taken alone, none of them provided conclusive evidence that coelacanths lived anywhere other than the western Indian Ocean; but taken together, at the very least they opened a doorway of possibility.

Among the most compelling was the Tampa scale—the unusual fish scale sent in 1949 by a Florida souvenir seller to Mr. Isaac Ginsburg at the Smithsonian Institute in Washington, D.C. At that point, the only coelacanth known to the world was Marjorie Courtenay-Latimer's, sitting stately and stuffed at the East London Museum, with all its scales accounted for. While the Tampa scale was not identical, it bore enough similarities to Latimeria chalumnae to convince Ginsburg that it was the scale of an ancient fish, probably a crossopterygian, and possibly a species of coelacanth.*

Though Ginsburg wrote to the woman asking for further details, she never replied, and the provenance of the Tampa scale remains a mystery—a tantalizing hint at the existence of an American coelacanth.

The next, probably even more compelling clue came to light in 1964, when an Argentinian chemist, Dr. Ladislao Reti, visited a small village church near Bilbao, on the Atlantic coast of Spain. There he saw a strange silver model of a fish hanging on the wall as an ex-voto—a prayer of thanks. It was about four inches long, and beautifully crafted, with the coelacanth's unmistakable paired, lobed fins, extra dorsal fin, and puppy dog tail, as well as an intracranial joint of its own, which unclipped to reveal the hollow center of the ornament. Every feature of the silver model was identical to the real thing, down to the engravings on the scales, which clearly represented the coelacanth's identifiable white markings.

Dr. Reti bought the model and took it home, where he gave it to an ichthyologist, Dr. Pablo Bardin, who identified it positively

---

*The scale is now lost, presumably somewhere in the vast museum archives.

Silver ex-voto of a coelacanth, found in Spain,
and believed to be of Mesoamerican origin.

as a coelacanth, and then to an American paleontologist who, after careful examination, concluded that it was probably of a different species than *Latimeria*. In appearance and detail—particularly the number of scales—it is more similar to the fossil genus *Macropoma*, the youngest of coelacanth fossils to have been discovered, dating back 70 million years.

Shortly afterwards, another even more exquisite silver coelacanth was bought privately by Maurice Steinert in Toledo, a short drive from Bilbao. Larger than the first ex-voto—almost exactly the same size as the Mozambique pups—it is even more delicately and intricately carved, undeniably a coelacanth. Steinert bought it from a collection of silver fish, probably the work of a single artist, and all of which depicted unusual fish with exceptional body shapes. In the opinion of an authority at the Prado Museum in Madrid on South American silver, it is of Mesoamerican origin, probably dating back to a seventeenth- or eighteenth-century Mesoamerican artist—certainly long before world attention was focused on the coelacanth. While Spanish silversmiths of that era stamped their work with the date and place of origin, Mayan Indian silversmiths were prohibited from doing so—and neither silver coelacanth had such a mark. It was common in those days for rich Spaniards to bring back works of art from their colonies in Central and South America, and donate them to the Church. Another silver expert who was shown the model by Plante and Fricke confirmed the Prado authority's dating: both the black silver oxides on the surface of the fish, he said, and the minute forged joint beneath the head (later such joints were cast, and appeared heavier) were consistent with seventeenth- and eighteenth-century work. Is it possible that these beautiful silver coelacanths are artistic representations of the same fish that shed the Tampa scale?

Hans Fricke and Raphael Plante were particularly fascinated by the question of how the silver coelacanths could have been crafted several centuries before the 1938 coelacanth was brought to the attention of the world. The representation of the white markings on the scales appeared to preclude a fossil muse, and anyway, the replica is so accurate in every tiny detail that it would have had to exceed the patient reconstructions of even the most skillful paleontologist. "For even the best of artisans to achieve such lifelike perfection from a fossil seems incredible," wrote Donald de Sylva in *Sea Frontiers.*

It is conceivable that the artisan was working from a fresh fish found in the Comoros long before it was identified by science, or brought back to Central America from the islands, salted or dried, but that again defies reason.

The easiest explanation is that the artists had live models to work from: a fish found nearer to home—somewhere in the waters of Central America. There are many remote ocean areas in the world, with similar environments to the Comoros—rocky, volcanic caves, and still, deep waters. Is it possible that they exist elsewhere in the fathomless oceans of the world—but that we don't employ the right methods to ensnare them?

Hans Fricke cannot discount the possibility. "I don't see why there might not be another population of coelacanths somewhere. I just hope we never find them."

XI

RAJAH LAUT

Indonesia

It was a dream wedding and honeymoon. Twenty-eight friends and relatives flew from around the world to Bali to witness American marine biologist Mark Erdmann marry his former research assistant, Arnaz Mehta, in a traditional ceremony. There was a procession of beautiful maidens carrying offerings of fruit and flowers on their heads, a wedding feast eaten from a long table groaning with local delicacies, dancing under a full moon. The next day, the wedding party boarded a tall-masted *phinisi perahu* for a six-day honeymoon cruise to the Komodo Islands.

At the end of the trip, most of the guests flew home, while Mark and Arnaz and two friends, John and Janel Intihar, continued their holiday in Sulawesi. For the Intihars it was an incredible experience—neither had been to Asia before. Their trip ended in Manado, a bustling seaside city on the northerly tip of Sulawesi, described by naturalist Alfred Wallace in 1859 as "one of the prettiest in the East." On the Intihars' last day, September 18, 1997, Mark and Arnaz took them on a final, cultural excursion to see a real, stinking Indonesian fish market.

It was a baking hot morning. As soon as they stepped out of the taxi, they were enveloped by the hustle and bustle of the market, people chattering and staring at them, everywhere the stench of fish. Arnaz caught sight of an old, wrinkled man wheeling a wooden cart with a scarlet A painted on the side, across the car park. On it, shimmering in the heat, lay a large and unfamiliar fish. She called to Mark and asked if he knew what it was.

"I immediately recognized it as a coelacanth," he recalls. "There was no doubt in my mind about it. I had read a book about the coelacanth when I was twelve, and it had really caught my imagination. At that time I knew they had only been found in the western Indian Ocean, but I wasn't up-to-date: I didn't know whether it had been found in this region since then, and I couldn't really believe that we had just stumbled into a major discovery."

Erdmann explained to the others what the fish was and why it was interesting. His mind was already racing as he tried to decide what to do, whether to buy the fish then and there. "I was rather wary of getting too excited about it, as I had been burned like that before. Apart from the eleven new species of stomatopods [mantis shrimps] that I've described from Indonesia, there have been all sorts of other creatures or types of behavior that I had thought were possibly new to science. I went to great lengths to examine them, and take photographs—sometimes even to the point of killing and pickling a creature in a jar. But inevitably, when I've taken them to the Smithsonian Institution in Washington or another such institution, I've found out that they were of no particular interest. It has been embarrassing, and

I've felt particularly bad about the animals I killed." The coelacanth, at least, was already dead; but on the other hand, it was far too large to fit in a jar.

Mark remembered that they were in town only for another few days, staying in a small hotel room, with a huge number of things to do. He and Arnaz were moving to Manado a month later, and had to arrange all the practicalities before they returned to America, so common sense won the day.

"I was really intrigued by it," Arnaz says. "I thought we definitely should try to find out more. A large crowd had gathered around us. I don't know what interested them more—us or the fish. They started speculating as to what it was, but they clearly didn't have a clue—most seemed to think it was a deepwater grouper." Arnaz encouraged Mark to photograph the fish and talk to the old man. He didn't have his own camera so he borrowed Janel's, and quickly took a few pictures of the coelacanth on its wooden cart.

While Mark examined the fish, Arnaz questioned the old man. "He seemed to be very uncomfortable with the attention," she recalls. "He looked as if he would be happier if we left him alone to get on with selling his fish to a market trader. I asked him where he had caught it.

"'*Dasar Laut,*' he replied. [On the sea floor.] This was not overly informative.

"I asked again: 'Where?'

"He turned and pointed out to sea, towards the offshore islands.

"'Far,' he said.

"'Do you catch them often?'

"'Rarely.'"

Mark continues, "His short answers stuck in my mind. As he was so disinclined to elucidate, I broke the first rule of questioning and started to ask leading questions. Anyway, when I pushed him, he said he had caught it in deep water, and seemed to suggest it was while he was hand-lining from his dugout. It was at night, and the fish hadn't wanted to die."

Mark Erdmann, in search of the Indonesian coelacanth.

The old man was becoming increasingly uncomfortable, and while they talked, his fish was baking in the hot morning sun. They decided to let him go about his business, confident that they would be able to recognize him again by the scarlet A on his cart.

"I remember my hesitation as he was walking away," says Mark. "Should I have bought it? Logic kept saying no. Of course, I soon came to regret my decision. I kicked myself about it for nearly a year; it was the biggest mistake I have ever made. I sat up many nights agonizing about it: at the very least I could have taken a scale, or blood and tissue samples—I even had my sampling kit with me, but it just didn't occur to me at the time. I consoled myself that we were going to be in Manado for two years, and we were bound to find another one."

Three days later, they flew to California. During the long flight, the coelacanth kept invading Mark's thoughts. The day he got to Berkeley, where he had recently completed his doctoral thesis on stomatopods, he went to see his department head, Dr. Roy Caldwell, and asked him if he knew whether coelacanths had ever been found outside the western Indian Ocean.

"Not to my knowledge," Caldwell replied. They searched the Internet and their reference books, but found no mention of coelacanths living within thousands of miles of north Sulawesi. They were both very excited about the apparent significance of the find, but Mark decided to wait until he had Janel's photographs from the market before getting in touch with any coelacanth specialists.

Four days later, he and Arnaz were in Ohio, visiting Mark's mother. They returned to the house after lunch to find that the phone had been ringing off the hook. Roy Caldwell had called several times, as had their friend John Intihar, and David Noakes, an ichthyologist from Guelph University in Canada. All, apparently, were excited about some fish, Mark's mother reported.

Mark contacted Roy, who said he possibly had made a big mistake. He had picked up an e-mail from John Intihar (a systems engineer in Washington, D.C.), which had been sent to everyone who had been at the wedding, urging them to visit "Mark and Arnaz's Honeymoon Website," which he and Janel had created. Roy, a self-confessed technophile, had immediately hooked up to the site, where he saw—among pages of wedding pictures—a photograph of the coelacanth. There it was, in full color, definitely and unmistakably Old Fourlegs. Roy's first excited thought was to e-mail some

Berkeley colleagues, telling them to sign on and "see what Erdmann caught in Indonesia."

Among the people he contacted was George Barlow, a well-known ichthyologist and fish behaviorist, who forwarded the message to David Noakes, who was a former student of his and a colleague of Eugene Balon's and fellow member of the Coelacanth Conservation Council. Noakes was well aware of the rumors of predatory Japanese aquariums and Chinese life-prolonging elixirs, and as soon as he saw the picture, he thought of the conservation implications. He phoned Roy to congratulate him on an amazing discovery, but warned him to get the picture off the Web immediately. A quick call to John Intihar, and the coelacanth picture disappeared within the hour. But by that time, a number of people had already learned of the Indonesian fish.

Roy's phone started to ring. There were calls of congratulations, but also—as is inevitable where the coelacanth is concerned—a great degree of skepticism. Among the first to contact him was Eugene Balon, keen to stress that it was without doubt a "honeymoon hoax." Vic Springer at the Smithsonian, initially skeptical, later came up with a theory that the coelacanth had possibly been caught by a Japanese trawler in the Comoros, but off-loaded in the Manado fish market on the way home out of fear of contravening the CITES regulations.

By the time Mark got back to Berkeley, he had decided not to publish his find right away. "I didn't want all sorts of people descending on Manado, and offering the fishermen large sums," he says. "I wanted first of all to dampen the skepticism by getting hold of another specimen for myself, then to make sure that conservation measures were in place before announcing it to the world."

The Erdmanns returned to Manado, where they had high hopes of finding another specimen before long. They moved into their house on the island of Bunaken, a fifteen-kilometer boat ride away, and one of the world's best dive sites. It was an idyllic place to live for two years, with magical coral gardens, sandy beaches, no motorcars, and a relaxed island population. They set about searching for a coelacanth. Their first stop was the fish market in Manado, where they spent days in the hot sun looking for the old man or his scarlet A cart. But there was no sign of either. The cart, they soon realized to their dismay, could have been repainted—and even if Mark and Arnaz found it, there was no guarantee it would lead them to the old man.

They tried another tack. Mark got reprints made of the photographs of the coelacanth and distributed them around the market, with the offer of 200,000 rupiah ($20) compensation to the person who brought him a coelacanth. None of the fish vendors, however, appeared to recognize the fish. Many of them said it was a grouper, or made other implausible suggestions. One man, however, appeared to know it. He said it was called *kabos laut*—roughly translated as "mudskipper of the sea"—which to Mark's ears sounded a definite possibility.

"By this time the search had really taken hold of me. I read everything I could find about the coelacanth, including *Old Fourlegs*, and the story gripped me from the first page to the last. Since I was a young boy, I have loved stories of marine exploration and discovery, and J.L.B. and the coelacanth adventure really appealed to me. I would have liked to have lived in Darwin's time, when instead of scientific specialists there were true naturalists who traveled the world discovering new and interesting things. So I was determined not to let this opportunity slip by: I was determined to find another coelacanth, even though *Nature* had indicated that they would be willing to publish a report based on the photographs alone. I didn't want simply to publish and kiss the coelacanth good-bye."

With the aid of a grant from the National Geographic Society, Mark decided to expand the perimeters of his quest. He began to tour the offshore islands, asking fishermen if they knew the coelacanth, and leaving copies of the photograph, with his address and the reward details on the back.

Rp 200,000—PER TAIL, MAXIMUM THREE FISH.
IF YOU CATCH ONE, PLEASE BRING IT QUICKLY AND
DIRECTLY TO BUNAKEN AND LOOK FOR
DR. MARK ERDMANN IN PANGALISANG BEACH.
PLEASE BRING IT IMMEDIATELY, BEFORE IT STARTS TO GO OFF.

He started on Bunaken. "The responses I got fell into three categories," he explains. "A blank stare; those who said they knew it as *buku laut* [literally, book of the sea], a big fish which they said shelters under flotsam during the northwest monsoon—I dismissed that as highly dubious; and those who said it was *ikan sede*. The last I found initially promising. They were mainly old hand-liners, who fish from pirogues identical to the Comoran *galawa*. They seemed to recognize the important aspects of the fish—the fins, tail, and scales—and said they typically found it at a depth of more than 100 meters, near the coral wall. They got very excited about catching it, and said they would have one for me soon. My hopes were raised."

The next week, he visited another small island, Pulau Nain, where he talked to shark net fishermen, who said they didn't know the fish. "You don't know *ikan sede*?" Mark asked.

"Sure, we know *ikan sede*, but this is not it," they replied.

"We went from high to low—it was an emotional roller coaster," Mark recalls. "Half the time I felt really discouraged, not knowing who to believe and who not to. Anyway, I left everyone with a photograph and my address."

In the second week of March, Mark and Arnaz visited the next-door island of Manado Tua, a defunct volcano that rises like a huge anthill out of the sea. The islanders live along the narrow fringe of beach; most earn their living from the sea, while a minority harvest coconuts, and farm bananas, mangoes, and hot bird's-eye chilies on the densely vegetated steep slopes of the mountain. Mark and Arnaz climbed to the top, leaving their cook, Tante Ita, whose late husband was from Manado Tua, to ask around among the fisherman. By the time they got down again, she was jumping with excitement. "I've found someone!" she cried, and took them to meet Om Lameh Sonathon.

Om Lameh, a slight man with a shy smile, was fifty-six, he told them, and had been fishing for fifteen years. Every evening, he takes his crew of eleven men on his flaking twenty-foot boat, *Trinitas*, to lay deep gill nets near the coral walls. The primary targets are deepwater grouper, snapper, and sharks, which have their fins

Manado Tua as seen from Manado harbor.

cut off before being sold to China and Taiwan for shark's fin soup. According to Mark, "He seemed to recognize the picture immediately. Yes, he had caught them—perhaps two or three a year—off the southeastern coast of Manado Tua. No, it was definitely not *ikan sede*—this fish was thicker and more oily. They called it *rajah laut*—king of the sea."

It was the first time Mark had felt confident about an identification. His confidence was further boosted when Om Lameh suggested he talk to another gill-net fisherman, Maxon Haniko, on the west coast of the island. "Maxon was a younger man, perhaps in his early thirties, with a rather cocksure manner. He also appeared to know the coelacanth, and was very keen on the prospect of looking for one for me, but unfortunately, he had a problem with his boat engine. . . . I agreed to lend him one of mine for a while. We went back to our boat to get it, and when I returned to give it to him, there was an old wrinkled man sitting with him, who looked vaguely familiar. Maxon introduced him as his father, and when I sat down and started telling my story again, about the fish in the market, the old man perked up and said excitedly, 'I was the man in the market!'"

This was the confirmation Mark was looking for. Maxon's father explained that he had been there selling a *rajah laut* caught by Maxon—he himself had never caught one while hand-lining from his small pirogue. He said he had sold it to a fish trader for Rp 25,000 (at that time, about $3), who had immediately sold it on to a Chinese person. In fact, the old man said, each time he had sold a *rajah laut*, it had been bought by a Chinese person. "He thought this was really funny," Mark recalls. "As a rule the Chinese here are more wealthy and more ostentatious with their wealth, and the fish has a reputation for causing diarrhea. The old man thought the Chinese bought it believing it was a grouper, and then ate it—to find themselves shitting oil!"

If the coelacanth was *rajah laut*, what, then, was *ikan sede*? Mark and Arnaz didn't have to wait long to find out. A few days later, their boatman, Daeng Said, came to the house early one morning to say that an *ikan sede* had been caught and was being saved for Mark by a fish broker on the beach at Bunaken village. Said didn't think it was the fish they were looking for, but they rushed to the beach nevertheless. There they saw a very large brown fish with big eyes, a big mouth, and short, spiny scales. Mark could see that it was a deepwater fish, but beyond its size, it bore no resemblance to the coelacanth. He thanked the fish broker and paid him a decent sum, took some photographs of the fish, and gave a portion each to Said and Tante Ita to try. Both were a little hesitant at first—they had heard the stories of oily diarrhea—but Tante Ita ate it, and pronounced it delicious.

Mark went home to try to identify *ikan sede*. He found a picture of the same fish on the U.S. Food and Drug Administration website. It turned out to be the coelacanth's old friend and Comoran neighbor, *nessa*—the oil fish, *Ruvettus pretiosus*. "We now knew that the Bunaken fishermen clearly didn't know what they were looking for—the tails of the coelacanth and the oil fish couldn't have been more different. We decided to concentrate our efforts on the Manado Tua crews."

Mark started visiting Om Lameh and Maxon every couple of

days. He raised the reward to Rp 600,000, roughly equal to the market price of two large sharks. "I thought carefully about the amount," he explains. "It had to be enough to make it worth their while to bring it to me if they caught one, instead of taking it to the market—but not so much that the news got out that there was this crazy guy who was offering a fortune for a fish, resulting in an armada of fishing boats descending on Manado Tua, all on the hunt for a *rajah laut*. I didn't want to cause excess mortality. At first I was concerned that I would be inundated with coelacanths— which is why I didn't immediately take the J.L.B. Smith route of plastering posters around entire villages. Instead, I tried to talk to the fishermen in likely places, and only when they showed signs of recognition, to offer the reward."

He soon realized he had been optimistic. The months passed and nothing came. Indonesia erupted into discord, the economy crashed, President Suharto resigned, and for a time the streets of most big cities swarmed with angry and violent protesters. North Sulawesi, however, remained quiet. Mark went out on Om Lameh's boat at dusk to see the fishermen carefully let down their 100-meter net along the steep reef wall, attaching stones along its bottom to make sure it didn't drift—and at dawn to watch them pull it up, a dozen wiry men working synchronistically, like a professional tug-of-war team. Each time the net surfaced, Mark expected to see a coelacanth caught in the mesh. Each time, he was disappointed. He studied the temperature and depth profiles of each fishing spot. It appeared that *rajah laut* lived at the same sort of depth and within the same temperature range (between 16 and 20 degrees centigrade) as the Comoran coelacanth. In the middle of July, he raised the reward to Rp 1 million ($100 in the devalued Indonesian currency). "I wished them good luck and told them to get on with it and bring me the fish."

On July 30, Mark and Arnaz Erdmann were waiting at home for their boatman, Said, to arrive with the boat to take them to the city.

"He was running late, which is unusual for Said," Mark recalls. "At about 8:10 I saw the boat coming around the corner. There were several people on board, which was a bit unusual, but I just figured Said had offered them a ride into Manado. He came running up the steps, and leaned—rather too casually, I thought—in the doorway of my study. He had this huge grin on his face, but he was determined to act cool. Arnaz said, 'Good morning. How are you?' at which point his cool broke. 'We have a *rajah laut!*' he said.

"I looked down and I could see this large fish being held by Om Lameh's son in the shallows. We raced down the steps, Arnaz with video camera in hand, to the beach, very excited. I had a closer look, and saw that, yes, it was definitely a coelacanth—the real thing! What happened then was very much like J.L.B. described. A million thoughts started racing around my head: what I should do, who should I tell?"

Arnaz filmed the coelacanth as it paddled slowly in a foot of water. It was clearly dying: it kept trying to roll over onto its back, and there was a tear in the delicate front dorsal fin. While she filmed Om Lameh holding the coelacanth in his arms, Mark took photographs. At no point could he wipe the enormous smile off his face. He told Said and Om Lameh to pose the fish for the camera: tail up, fins out. But all the time, his mind was swirling with all he had to do.

"As the fish still looked alive—just—we decided to take it into deeper water for some more photographs. We grabbed our diving gear and my underwater camera, and took the coelacanth out to the reef flat, where we photographed it at a depth of two meters. However, as the visibility wasn't great, and it wasn't its natural habitat, I suggested we take it to the reef edge. Arnaz was against it. The fish was bleeding slightly and she was worried it might be grabbed by a shark. But I felt ready to do battle with any shark."

Mark managed to convince her, and they towed the coelacanth behind the boat across the reef. As the water passed over its gills, it appeared to revive a little; it stopped trying to turn belly up and started to scull with its fins. They dived into the water with it,

Arnaz Mehta Erdmann swimming with the coelacanth.

and guided it down a few more meters. It offered no resistance, and didn't try to struggle or escape.*

"We were down for about forty-five minutes. There was a strong current and bad visibility," Mark recounts. "I was very concerned about sharks, and kept looking around for them. I was also very caught up in the technical details of positioning it right for the camera, and trying to keep Arnaz out of the picture."

"I swam beside the coelacanth with the rope in my hand," Arnaz continues. "Every now and then it tried to roll over, so I would support it and put it back into a swimming position. I was struck by its beauty—it looked like it was cloaked in golden armor. It swam in a very unhurried manner—it didn't even appear to be scared. Once

---

*As they swam along the reef wall, a dive boat passed over them. Coincidentally, the man inside was Peter Scoones, the BBC cameraman who had filmed the dying coelacanth in the Comoros in 1977. This time, however, he was filming spawning clown fish, and remained oblivious to the presence of another coelacanth.

in a while it would take a big gulp of water. It reminded me of a Spanish dancer, waving its fins like a flouncing skirt."

"It was definitely feeling a bit better by that stage," says Mark. "When I moved in for some close-ups, I really had a chance to take it in. It was magnificent, each scale appeared to be flecked with gold. I touched it and it was very soft: I could put my arms around it and squeeze, and it was more like holding a baby with soft, young flesh, than a big, hard fish. The thing that captivated me most was its eyes. They were large, and in certain lights were a luminescent, almost alien green, and they kept looking at me: wherever I went, its eyes followed. When we were in the shallows, the fisherman had said: 'Look out for its mouth—don't let it bite you,' but the thought never crossed my mind. It seemed very gentle and calm."

They finished the film, then took the coelacanth back to the reef flat. Mark grabbed his dissecting kit, sample vials, liquid nitrogen container, and alcohol, then they put the fish in some water in a big green cooler box and loaded it onto the boat. It was the largest chest they could find, but it was still too small for the coelacanth, which lay rather uncomfortably, its unusual tail poking out of the end, flapping its fins feebly from time to time.

"I was filled with excitement and adrenaline, but at the same time it was heartbreaking to see it slowly dying, especially having swum with it," Mark recalls. "It seemed so special, there was no hint of ferocity. At risk of slipping into anthropomorphisms, I had the impression of great gentleness and intelligence. I used to do a lot of spear fishing, and I have seen many dying fish—most of them are anything but dignified. They shake and thrash around. But the coelacanth seemed to me to be very stately. It was very, very sad. I can honestly say that if it had looked more alive when we had been photographing it, I would have had the impulse to let it go."

J.L.B. Smith would have appreciated Mark Erdmann's dilemma; yet he, too, would have made the same decision. Mark's swimming companion doubtless would have died within hours of its release—to this day, no coelacanth has survived the trauma of capture—and its precious inner secrets would have been swallowed in an instant by the predatory sharks that patrol the the outer reef.

Those secrets will be the start of a new chapter in the study of one of the world's most miraculous creatures, and may contribute to finding a way to ensure its continued existence for another 400 million years.

The coelacanth remained alive for most of the half-hour boat journey to Manado, moving less and less as time passed. Eventually, only its large green eyes showed any sign of life, and as the boat approached the harbor, they too became still, as quietly, with dignity, the Indonesian king of the sea died.

XII

TERRA INCOGNITA

On September 24, 1998, Mark Erdmann's article was published in *Nature*, as J.L.B. Smith's had been almost sixty years earlier. It was immediately dubbed "the zoological sensation of the decade" and accorded an appropriate reception. Across the world, the coelacanth's picture was splashed across newspapers and onto television and the Internet. "Second Home of Fish from Dinosaur Age Is Found," declared the headline in the *New York Times* above a long article that described the coelacanth as "ugly but fascinating." "Fish on the slab was 360 million years old," explained the *Daily Telegraph*, while the news channels—CNN, BBC, ABC, and FOX—concentrated on the human story of the Erdmanns' honeymoon adventure. Jerry Hamlin's dinofish.com website received a record 6,500 hits. It was a far cry from Mr. Adams of the *Daily Dispatch*, who had broken the story of *Latimeria chalumnae* in East London in 1939.

The coelacanth world, predictably, was thrown into a state of near hysteria. Certain scientists claimed to have been party to the secret all along, while others forgot their initial skepticism and reached for their ocean current charts. Cyber-lines between the major ichthyological centers buzzed as coelacanth specialists adjusted their assumptions, ate their words, and furiously debated the implications of the announcement.

Within minutes of the news reaching him, Hans Fricke was contemplating taking *Jago* to Manado Tua. His first feelings on hearing of a new population of coelacanths were mixed. "It is great, fantastic news for science," he said with a big smile. "It seems as if Old Fourlegs is tougher than we thought." He was, however, struck by the possible effects: "I feel sorry for the Comoran people. I hope it doesn't affect the conservation effort there. Maybe the coelacanth has a stronger chance of surviving in Indonesia, where they can learn from the mistakes made in the Comoros, and start trying to protect the fish from the beginning."

For the majority of the scientists, however, Mark Erdmann's discovery of an Indonesian chapter was something to cheer about. Not only did it open up new areas of research and provide them with fresh topics of discussion, but it unequivocally indicated a larger world coelacanth population than previously had been believed.

"It is just incredible that they were found in Indonesia, so far from where we had suspected the ancestral habitat to be in Madagascar or the Comoros," says Robin Stobbs. "It appears that both populations of coelacanths live at similar depths, at the same temperatures, in similar rocky and volcanic habitats, and together with the oil fish. Both are caught around the new moon (dark nights), and the deep shark nets used by Indonesian fishermen are identical to those used by ZeZe, the Malagasy fisherman who caught both the Madagascan specimens. The historical parallels, as well as the similarities in habitat, are remarkable. The Comoros and Madagascar were first settled by Indonesians nearly two thousand years ago, who brought with them their fishing skills and techniques."

More than six thousand miles separate Manado from Moroni, comprising the vast belly of the Indian Ocean. Two millennia ago, strong men in longboats set sail from their Indonesian home. Riding the currents, they headed west into terra incognita. Eventually, the lucky ones landed on the wild and unpopulated islands off the African coast. Perhaps they found there, among the strange fauna and flora, a large, scaled, and familiar fish?

To Mark Erdmann's first glance, the Indonesian coelacanth appeared to be almost identical to its Comoran counterpart. The only noticeable difference was the color: while live Comoran fish have always been described as steely blue with white splodges, the *rajah laut* was distinctly brown—with the same white markings, but also an intense glittering pattern of gold flecks covering its flanks. The golden shimmer, the prismatic effect of light reflected off the small, spiny denticles on its scales, had never been noted before, and could be taken as an indication, Mark believed, that his fish was not an exact replica of the Comoran *Latimeria chalumnae*. The proof would be determined by detailed genetic examination.

Barely an hour after his fish had died, Mark Erdmann had taken tissue samples of its major organs and stored them in liquid nitrogen. Two months later—four days after the *Nature* article appeared—he was joined by Susan Jewett, a curator at the Smithsonian Institution and fellow coelacanth enthusiast. She had come to Indonesia to help Mark prepare and preserve his coelacanth. Together they took it from his landlady's freezer—its temporary home—carefully packed it into a polystyrene coffin, and escorted it by plane to Jakarta. From there, they took it by car to the Bogor Zoological Museum, an hour's drive away. It was rushed into the museum's sparkling dissection room, where an audience comprising the cream of Indonesia's scientific community had gathered. Mark and Susan donned long gauntlets and gas masks—protection against formalin fumes—as if ready to engage in biological warfare. They weighed (66 pounds/29.2 kilograms) and measured (49 inches/124 centimeters) the fish, then injected it with preserving fluid, and arranged it in a lifelike position for exhibition. While they were preparing it, they found three small eggs in its stomach, a clear indication that "it" was a "she."

By this time, a rumor had emerged in coelacanth circles that Susan Jewett's express purpose was to smuggle the four-foot fish back to America in her suitcase. This, Mark was quick to deny. "The first fish was always going to be Indonesia's," he explained. "Hopefully we will be able to arrange a CITES permit to send the next one to Washington, which is exactly halfway around the world from Jakarta."

Susan Jewett returned to the Smithsonian to wait, her suitcase containing only the clothes she had arrived with. But by the time Mark returned to Manado, coelacanth mania had set in. The national media were drunk on the good news story, and the now predictable collection of rumors and strange tales had started to emerge. A local man had been overheard in the fish market offering Rp 2,000,000 for a *rajah laut*. An American scientist, desperate for some fresh coelacanth brain tissue, apparently was organizing an expedition to the islands. An Indonesian dive operator was insisting that *he* had found the coelacanth while diving, and a regular visitor to a guest house on Bunaken rather improbably insisted that the owner frequently cooked and served coelacanth *satay* to his guests.

The French, inevitably, tried to trump Mark Erdmann's discovery. Within hours of the *Nature* announcement, a cryptozoological website was claiming that in 1995 a French fishery consultant named Georges Serres had caught a 22-pound (10-kilogram) coelacanth south of Java while trying to fish for lobsters at night. He had salted and dried it before giving it to the local fishery department, who promised to send it to the Oceanographic Institute in Jakarta. Inconveniently for Serres, all his belongings—including photographs of the fish, known locally as *ikan formar*—were stolen as he was about to leave the country, and the institute apparently has no record of having received the specimen. "If we could find someone who could verify that the specimen prepared by Monsieur Serres was still at the Institute in Jakarta, it would be a pretty counter-scoop with a nice nationalistic flavor," wrote one French scientist.

What was more worrisome, in the month following the announcement of the Indonesian coelacanth's whereabouts, the fishermen of Manado Tua—local superstars Om Lameh and Maxon and their crews—were approached by at least five groups of Japanese (or perhaps the same group five times), including a persistent delegation from the Toba Aquarium of Tokyo, who had failed in their earlier multimillion-dollar effort to catch a live coelacanth in the Comoros in the late 1980s. These groups offered the fishermen ever-increasing sums in an attempt to persuade them to cooperate in special coelacanth-hunting expeditions.

Mark Erdmann, however, is determined to prevent anyone from poaching a coelacanth from under his nose. Like so many people who have come into contact with it, he has fallen under the coelacanth's spell. Since September 18, 1997, when he saw his first fish in the Manado fish market, he has thought and dreamed of little else.

At the end of 1998 and throughout 1999, he made several trips to Jakarta to meet with the other members of a hastily convened coelacanth think tank. The outcomes were upbeat: it was agreed that there would be a ministerial decree declaring the coelacanth to be part of Indonesia's national heritage, and as such to be protected for future generations. Attempts to target the coelacanth for capture were outlawed for the present, and it was decided to print and distribute throughout the fishing villages—first of Sulawesi and later all Indonesia—a poster describing the coelacanth, its conservation status and CITES listing, and urging that any catches be reported directly to the nearest authority—for which a small, unspecified reward was offered. The proposal that there should be a coelacanth information center was also greeted with enthusiasm.

"The meetings have been well run, well attended, and lively," reports Mark Erdmann. "Both national and local awareness of the coelacanth have grown tremendously—helped by excellent local media coverage. Perhaps most excitingly, the villagers are beginning to benefit from the coelacanth. Arnaz and I have been working with a woman from Bunaken who is extremely talented at embroidery, and she is now producing coelacanth T-shirts, hand towels, and handkerchiefs, and doing a brisk trade with tourists. She has begun to train additional village women to make these wonderful coelacanth souvenirs. If local people can benefit from the coelacanth, there can be no better way to ensure its protection."

*   *   *

In early 1999, the first results of the important DNA investigations came back from the laboratory. (The tests compared the DNA in the tiny mitochondria—thousands of bacteria-like power plants—in each cell, rather than the more complex DNA in the nucleus.) A sequence of 3,221 bases of the mitichondrial genetic blueprint of the Indonesian coelacanth was obtained, and after extensive massaging through a computer, it was calculated that they showed a 4.1 percent divergence from the Comoran coelacanth's sequence. This is a substantial difference and probably enough, in Mark Erdmann's opinion, to call it a new species. (In the salamander world, such a divergence would indicate two separate species; while in bird and snail terms, it is well within the range of variation of a single species.) He wanted to wait for the results of detailed morphological comparisons between the Indonesian and the Comoran fish before making a final decision.

Others were not so cautious. In March 1999, just as the Erdmann team's results were awaiting publication in *Nature*, the French finally got their patriotic counterscoop when a French catfish specialist, Laurent Pouyaud, reported the results of his Franco-Indonesian team's DNA investigations in *Comptes Rendus de l'Académie des Sciences*. From his results, Pouyaud appeared to have no hesitation in naming the Indonesian coelacanth as a new species, *Latimeria menadoensis* L. Pouyaud. Mark Erdmann was incensed at what he saw as a piece of scientific piracy on the part of the Frenchman, and a mighty row broke out in the coelacanth world, of a magnitude only surpassed, perhaps, by the very public front-page fight following J.L.B. Smith's "theft" of the second coelacanth in 1952. Systematicians were particularly condemning of the ethics of the enterprise. There was no reference to the discoverer of the specimen (Mark Erdmann—well known to be studying it), no designation of the holotype, no museum number, and a widespread publication of the new name in the press before the scientific publication appeared. Nonetheless, it was deemed to conform to the International Code of Zoological Nomenclature, and, as such, is scientifically and legally acceptable. In his defense, Pouyaud explained that he had been urged to name the fish by his Indonesian colleagues.

Both scientific teams attempted to estimate the date when the two coelacanth "families" parted ways. This was always going to be a best guess. As the Comoran *Latimeria chalumnae* has no close relatives, it is impossible accurately to determine when the genes split; but Erdmann's estimates put it between 5.5 million and 7.5 million years ago, while Pouyaud suggested a few million years later.

Either way, their results appear to answer the conundrum that has been exercising the minds of the coelacanth junkies since the existence of the Indonesian fish was announced: which came first, *gombessa* or *rajah laut*? Seven and a half million years ago, there were no Comoro Islands, the volcanoes had yet to erupt out of the sea to create land. Great tectonic movements were still changing the face of the earth. It is possible, Mark Erdmann suggested, that the massive plate movements that led to the formation of the Indo-Australian Arc and the resultant separation of the Indian and Pacific Oceans in the Miocene period (from 25 million years ago) cut a part of the coelacanth population off from the rest. If this is what happened, it is conceivable that there are more coelacanths, in different locations across the globe. "It seems highly unlikely that the living coelacanth exists only in two small, highly disjunct populations," Mark Erdmann wrote in *Nature*. Fired by the enthusiasm of the response, he was eager to start showing his photographs in coastal villages and fish markets around the Indonesian archipelago.

The possibility that there might be coelacanths off other exotic islands undoubtedly will be spur enough for a whole new generation of adventurers and enthusiasts to print up their own reward posters and start searching the more remote fringes of the Indian Ocean. The suggestion that the two silver coelacanth models might have been of Philippine rather than Mesoamerican origin, and carried back from the Philippines (only a few hundred miles from Manado) by Spanish merchant traders to Toledo and Bilbao, adds fuel to their fires.

The rumors surrounding the coelacanth doubtless will continue to attract enthusiasts and eccentrics, or perhaps to drive its admirers to eccentricity. A few years ago, the daughter of a German count sent Hans Fricke a beautifully handwritten treatise the length of a book, extolling her theory of how and why the silver coelacanths were made. She had acquired this knowledge, she explained, from free-floating electromagnetic rays. An East London taxi driver, C. D. Harrald, has spent his entire life telling people that he was the man who nearly refused to take Marjorie Courtenay-Latimer, her assistant Enoch, and "that stinking fish" from the docks to the museum on December 22, 1938. What he fails to recall, however, was that he was too young to have had a license at that time, and that it was his uncle—who died in 1969—who was in fact driving the taxi.

Even when the coelacanth was known only as a fossil, it captured the imagination, and it appears to have maintained its allure. In the months after the existence of an Indonesian population became public knowledge, numerous scientific and photographic expeditions were being planned, and in late 1999, Mark Erdmann joined Hans Fricke and his team of "jagonauts" in a sadly luckless exploration of the Indonesian islands.

"It was an amazing experience," he said. "I bet J.L.B. Smith would have loved to have seen the natural environment of the coelacanth." Without doubt, coelacanths will be found in Indonesia, photographs will be taken, films will be made, and in time, the hidden world of *rajah laut*, swimming below the reefs of Manado Tua, will be revealed to us. The preserved Indonesian coelacanths will sit alongside their Comoran cousins in the museums of the world. With any luck, many will continue to evade our nets and lines. "The fact that living coelacanths could escape detection in an area well studied by ichthyologists for over a hundred years is wonderful," Mark enthuses. "It is a humbling and exciting reminder that humans have by no means conquered the oceans, and provides hope that Old Fourlegs is more abundant and resilient than we initially dared hope."

Human beings must appear mere parvenus in the ledgers of coelacanth history. Our important events, from learning to make stone tools to walking on the moon, are but a small bleep to the silent witness at the bottom of what we presume to call "our" oceans. It is comforting to imagine the coelacanth swimming quietly around, watching all of the crazy things that take place, surviving far greater

Marjorie Courtney-Latimer accepting her gold
coelacanth coin, Cape Town, South Africa, 1998.

holocausts than those we have known, and continuing to exist after
this extraordinary duration of time.

On July 31, 1998, at the very same time as the gold-flecked
Indonesian coelacanth was swimming into Om Lameh's net, six
thousand miles away, Marjorie Courtenay-Latimer was the guest of

honor at a ceremony organized by the South African Mint to launch a limited collector's edition of gold coelacanth coins. The evening was hosted by the Two Oceans Aquarium in Cape Town. Marjorie, a sprightly ninety-one years old, was introduced by Mike Bruton: "Ladies and gentleman, this evening we are in the presence of one of the great figures in South African and Western science," he began. "Someone who played a vital role in the greatest biological discovery of the century, the discovery of the first living coelacanth. . . . I refer of course, to Marjorie Courtenay-Latimer."

Dressed in a neat black suit with a fake fur collar (which, she joked, was almost as old as the coelacanth), Marjorie approached the microphone to deliver a short speech. She wanted to thank the South African Mint, she said, for the great honor, and also to trace briefly the events that led to her discovery. As she spoke about Captain Goosen, Bird Island, and J.L.B. Smith, about seeing the beautiful blue fish for the first time and her absolute conviction that it had to be saved, she put down her prepared notes. Speaking only from her sharp memory, she transported the audience back sixty years to a small museum in East London, and a young woman who was determined that the strange fish she had found was something special and had to be saved.

"I'll never stop on the coelacanth," she said after the presentation. "It's like perpetual motion. Sometimes I think I am sick to death of it, because there are so many other things I am interested in. I spent forty years at the museum and built it up from scratch, and it was about so much more than the coelacanth: my aim was to make it so that, if you only had fifteen minutes to spend, you would be able to enjoy it. I arranged all the fish with pools of water, corals and seaweed; all the birds and animals in natural settings. When I retired I went to live in Tsitsikamma for fifteen years, on a little farm, and wrote a book about the wildflowers in the national park down there. But I could never completely leave the coelacanth behind. On the fiftieth anniversary of its discovery, I was invited to the Comoros for a special ceremony at the museum. It was a wonderful experience: really I have been a very privileged person, a very spoilt person because of the coelacanth. The coelacanth put me on the world map, and I still get letters from all over—and lovely letters from schoolchildren—asking about the coelacanth. I try to reply to every one of them. And without the coelacanth, of course, I wouldn't have been here tonight to receive this lovely gold coin."

Two months later, when the existence of the Indonesian coelacanth was publicly announced, Marjorie Courtenay-Latimer was once again swamped by journalists and broadcasters eager to hear what she thought about the latest chapter in the coelacanth story. "I think it is wonderful news, very exciting," she said. "To think that this started so many years ago, with Captain Goosen in East London, and my desperate search to try to find a way to preserve the strange and beautiful fish he had caught. . . . It really is a remarkable creature, and I hope that the new interest in it will prompt us to find a way to save it, once and for all."

APPENDIX A

Throughout this book, I have attempted to describe and explain as much of the science of the coelacanth as I judged necessary for readers to feel they had been properly introduced to this most enigmatic of creatures. In the interests of an uninterrupted narrative, I have included these details in their chronological setting: what J.L.B. Smith discovered through his examinations of the first and second specimens is covered in Chapters 2 and 5; Millot and Anthony's work in 6 and 7; Hans Fricke's observations on living creatures in 9 and 10; new revelations about the reproductive system in 10; and recent DNA and RNA analysis of coelacanth blood and tissues in Chapter 12.

As the more technical information is scattered throughout the pages, I thought some readers might be interested in a more in-depth, consolidated look at the science of the coelacanth. This is not meant in any way to be exhaustive. Those who want to research further can find inspiration from the reading list which, I must stress, is only the tip of the enormous iceberg of writing on the coelacanth that has been published since Louis Agassiz identified the first fossil in 1839, and that continues to be published at an undiminishing rate.

There are 26,000 species of living fish, but few have captured the public imagination in the way that the living coelacanth did when it was revealed to the world sixty years ago. This strange survivor continues to fascinate; it barely has to flap a fin to merit a photograph and a decent-sized story in national newspapers around the world. Undoubtedly, its appeal is closely connected to the romantic idea that it is the "missing link" between the marine and terrestrial world: the first fish to crawl out of the sea to conquer the land. While its exact relationship to early tetrapods is still a matter of furious debate and little agreement, what is at least as interesting—and even less known—is the coelacanth's counter-evolutionary role. Why has this ancient lineage survived, while its contemporaries either perished or evolved?

Despite millions of man-hours of research, during which every organ of the coelacanth has been exhaustively dissected and analyzed, there is an enormous amount still to learn. The coelacanth is like a patchwork quilt: we have the squares of exquisitely embroidered cloth, but not enough thread to sew them together.

To stretch the metaphor: this thread is probably only to be found in haberdasheries that are yet to open. Scientists do not have the tools to unravel all of the mysteries of the coelacanth. In order to learn much more, a living coelacanth must be studied in detail—something which, until we can observe its behavior for long periods in the wild, or until we have more sophisticated methods of examining it in our own world, either dead or alive, is still some way into the future.

In the pages that follow, I will look at the anatomy, physiology and behavior of the Comoran coelacanth, *Latimeria chalumnae*. At time of press, the Indonesian coelacanth had been named as a new species, *Latimeria menadoensis*, but as yet there had been little in the way of detailed morphological evidence to back this up.

Fins and Scales

In appearance, at least, the coelacanth is less a fish than a bizarre confection of mismatched parts—modern, ancient, and unique. As one of the deckhands on Captain Goosen's ship, the *Nerine*, said after hauling it up off the southern African coast in 1938, "It looks like a giant sea lizard." Its color, perhaps, reinforces that image: blue-gray with irregular white blotches, its scales are hard and closely interwoven, the tiny pattern of denticles rough to the touch.

The coelacanth's fins are perhaps its most distinctive feature. The paired pectoral and pelvic fins—the skeletons of which bear a marked similarity to those of lizard legs, and indeed, to our own appendages—appear to be attached to fleshy, limb-like peduncles, as do the unusual lobed rear dorsal and anal fins. The front dorsal fin (apart from sharks, all other fish have one dorsal fin only) is delicate and fan-like, similar to that of the vast majority of fish. The tail, in contrast, is completely different: it is divided into three sections, the diamond-shaped trunk being intersected by a small, supplementary epicaudal fin, memorably described by Marjorie Courtenay-Latimer as its "puppy dog tail."

Hans Fricke's films show the coelacanth's swimming action: most of the time, it drifts slowly with the currents, using the extremely mobile pectoral and pelvic fins in a scull-like fashion to control direction. Like our shoulders and hips, the pectorals can be rotated through 180 degrees. Initially, it looks as if the coelacanth's many fins are moving seemingly at random, but on closer study, it is clear that the two pairs of fins are being used in a diagonal fashion: right pectoral and left pelvic moving together, left pectoral and right pelvic, rather like the gait of a trotting horse, or that of lizards and many other tetrapods. When startled, the coelacanth propels itself forward with its strong caudal fin and a twist of its body; it presumably uses the same short, sharp bursts of speed to catch its prey. It has also been observed swimming on its back, and of course, standing on its head.

The Head

The coelacanth's tiny brain lies entirely behind a unique intracranial joint, a feature previously known only from fossils of primitive fish. It was not known at first if the coelacanth used this joint, until Keith Thomson of Yale University examined their specimen in 1966: "The fish was completely thawed and flexible. I grasped the tip of the snout (carefully because the teeth are sharp) and lifted. . . . The joint hinged smoothly, the tip of the snout came up, and the lower jaw dropped and moved forward. . . . I felt a little like Galileo." This enables the coelacanth to increase its gape by a sudden opening of the whole head, giving it a powerful bite that compensates for its relatively small teeth. The slow-moving relic thus revealed itself to be a powerful predator.

The Rostral Organ

The rostral organ, a large, jelly-filled cavity with three exterior openings on either side of the snout, is also unknown in living creatures (although it was described in the snout of fossil coelacanths). It was initially thought to have something to do with the respiratory system, but more recently has been thought of as some sort of electro-detector. The coelacanth, it is hypothesized, is able to sense the presence of predators or prey in the dark by "hearing" the tiny impulses traveling down their nerves and muscles—even the beating of their hearts. The core of the earth emits electromagnetic rays much as the sun does light, and it is possible that coelacanths can thus "see" their night landscape as clearly as we see ours in daylight. When Hans Fricke experimented by emitting a simple electrical pulse from his submersible, he observed that the coelacanth frequently raised itself into the headstanding position to bring the snout close to the electrical source. He has proposed that this might be connected with the electroreceptive function of the rostral organ, and that the coelacanth might have, in addition, further sensors in its tail.

The Eye

The eyes are clearly adapted to living in the gloomy depths of the ocean. They are fairly large with a preponderance of rods (sensitive to faint light), and virtually no cones (which recognize color—meaning the coelacanth has very poor color vision). There is no melanin in the retina, indicating a great degree of photosensitivity.*

*The eye has a tapetum—a reflective layer much like that of a cat—behind the retina, which acts as a mirror, so that light is intensified, thus allowing coelacanths to see very well in the dark—and conversely they are blinded in the light.

While Comoran night fishermen describe the eyes of recently caught fish as "shining like burning coals," in daylight, they quickly become an opaque, milky-green.

The Ear

One of the most interesting features is the coelacanth's inner ear. In 1987, Bernd Fritzch reported the presence of a basilar papilla—a thin membrane supporting a sensory epithelium, which is present in some amphibians and mammals and is only found in animals that live in air. This, Fritzch concluded, is a sign that the coelacanth was developing some of the prerequisites of land life—perhaps because it was spending at least part of its time above the water—and suggests that *Latimeria* and tetrapods are more closely related than previously believed.

The Skeleton

The skeleton, however, provides contrary evidence. It is believed that the first vertebrates to live on land had a fully developed skeleton, composed of bone. The coelacanth's, however, is made up predominantly of flexible cartilage, more similar to that of sharks, or other ancient creatures. Instead of a spinal column, it has a thick-walled, fibrous notochord, which is unsegmented, elastic, and filled with the viscous golden oil rumored to act as a life-prolonging elixir. Most vertebrate embryos start with notochords, but quickly replace them with vertebrae.

The Swim Bladder

The coelacanth's swim bladder is a slender oil-filled tube embedded in fat below the notochord, opening into the dorsal side of the gut. In most fish, the equivalent feature—usually known as the air bladder—is filled not with oil but with air, and it is easy to imagine how this came to be regarded as the precursor of the lung. While wholly different from these hollow organs of bony fishes—it certainly could not function as a lung—the coelacanth's fat-filled swim bladder similarly serves to increase its buoyancy.

It was widely held that one of the causes of coelacanths always dying soon after capture was decompression. In most deepwater fishes, the air-filled bladder literally explodes as the water pressure drops when they are raised to the surface. However, recent work by Hans Fricke, among others, shows that in coelacanths this is impossible: there is no air or gas anywhere inside the coelacanth, and oil does not significantly change volume or density in the face of changes in pressure.

Blood

The inability of the coelacanth to survive on the surface probably has more to do with its blood. The amount of oxygen which the blood can absorb from the water via the gills is highly susceptible to temperature changes. The results of scientific analysis of the blood suggest that the coelacanth must remain within the relatively low temperature range of 13 to 20 degrees centigrade in order to avoid suffocation. In warmer water near the surface, it would find it increasingly hard to extract sufficient oxygen.

The blood of the Yale specimen was also examined in 1966 by a colleague of Thomson's, Grace Pickford. Thomson describes her returning from the lab: "Grace, looking shocked, came back down to where we were working and announced, 'The blood is isotonic with seawater. It is full of nonprotein nitrogen . . . it is the blood of a shark.'" This high level of urea in the coelacanth's blood enables it to control its osmotic balance. In most fish, the concentration of basic blood chemicals is roughly one-third of the concentration of seawater—thus most marine fish have to cope with losing their body water to the sea by osmosis. So, bizarrely, they are involved in a constant battle to avoid dehydration. Sharks and rays fight this battle by retaining huge quantities of urea—created in the liver by a combination of waste ammonia and carbon dioxide—until their blood has the same osmotic concentration as seawater. Coelacanths retain enough urea to take their blood to 95 percent of the necessary concentration, and it is assumed that they make up the difference by ingesting small quantities of seawater—in the same way, though on a smaller scale, that most bony fish deal with the problem.

DNA

An analysis of the DNA sequences of coelacanth blood was first carried out by a German team comprising Thomas Gorr, Traute Kleinschmidt, and Hans Fricke. They determined the sequence of the α- and β-globin chains of coelacanth hemoglobins, and compared them with those of bony and cartilaginous fish, tadpoles, and adult amphibians. They suggested that both *Latimeria* chains were more similar to those of tadpoles than were any other bony fish. In fact, the coelacanth showed a significantly closer match than the lungfish. "Thus the primary structure of *Latimeria* haemoglobin indicates that the coelacanth is the closest living relative of tetrapods," the team concluded.

There are, however, several different methods of measuring relationships. The study of cladistics—a comparison of uniquely shared characteristics of different creatures—is currently accorded much weight, and this system clearly shows that the lungfish is more closely related to tetrapods than the coelacanth. Until such time as we have a *Back to the Future*–type vehicle, or we develop an incontrovertible method of deciphering these relationships, the coelacanth's place on the evolutionary tree will remain uncertain.

Sex and Growth Rate

There are no obvious external differences between male and female coelacanths, although fully grown females do tend to be larger. To date there has been no record of a male over 1.65 meters. The longest recorded coelacanth was measured at 1.8 meters—almost six feet—and weighed 95 kilograms. The pregnant Mozambique female, with her twenty-six nearly full grown pups, weighed in at 98 kilograms, and was a mere centimeter shorter at 1.78 meters. The life span of a coelacanth is thought to be as long as fortry years. Its growth rate can only be estimated: Mike Bruton puts a figure of about 6.5 centimeters per year, slowing later in life, until it achieves a length of about 170 centimeters after twenty years. The smallest recorded fish, a female juvenile caught in Iconi in 1974, weighed less than 1 kilogram and measured 42.5 centimeters.

While several juveniles have been caught off the Comoros, they have never been observed underwater. Fricke believes that they might live at greater depth—below the 400-meter threshold of his submersible.

Feeding Habits

We have yet to see a coelacanth feed, but examinations of the stomach contents of caught specimens indicate that they are predominantly fish-eaters. Their prey includes squid, lantern fish, cardinal fish, eels, skate, sharks, bream and beardfish—which sometimes reach sizes of up to three feet and are swallowed whole. They appear to shelter in dark rocky caves by day—when the water temperature is higher—and hunt at night. Fishing records suggest that they ascend a few hundred meters, where prey is more plentiful.

The coelacanth has one of the slowest metabolisms of any fish. This means it has to conserve energy, which it does by drifting around slowly, but it also is able to go for long periods without feeding. According to the local fishermen, they are never caught at full moon, and seem to be most prevalent on completely dark nights—when they are in no danger of damaging their highly photosensitive eyes.

Reproduction

The hows, wheres, and whens of coelacanth reproduction have been the subject of considerable debate over the decades, and much remains unknown. What is now sure is that the coelacanth is ovoviviparous—it is fertilized internally (although how is still unsure, as the male lacks an obvious intromittant organ) and gives birth to live young. Evidence suggests that only one ovary is functional, normally the right, where up to twenty-six huge, soft-shelled eggs, about 9 centimeters in diameter weighing and weighing 320 grams, develop. The yolky insides of the eggs provide nutrition for the growing fetuses during a gestation period of indeterminate length. By the time of birth, most of the yolk has disappeared, and the pups, around 32 centimeters in length, resemble miniature adults. The Mozambique fish, with twenty-six pups of around that size still in her uterus, was carrying an estimated extra 12 kilograms of weight.

Little is known about the age or size of sexual maturity, the method, place, and time of mating, or where the young are born.

With so many blanks in our knowledge about coelacanth reproduction, it is hard accurately to predict the effects of human predation on its population size and thus the chances of the future survival of the species. Inevitably, the debate over whether we should capture a live coelacanth will run and run, with the proponents stressing the imperative of studying a coelacanth up close, and the opponents arguing that the species would have a better chance of survival if the fish were left alone—much as they have been for 400 million years.

What is certain is, while there is still so much unknown about this most enigmatic of fish, we will continue to be fascinated by it. In 1990, Peter Forey wrote that coelacanths "have achieved a reputation that only few earn, but most never attain." Or more simply, as a schoolchild, in response to the question posed in a German magazine article, "Why it is worthwhile living this week?" replied that ". . . coelacanths still exist."

APPENDIX B

WHERE TO SEE
A COELACANTH

The Coelacanth Conservation Council (CCC) maintains an inventory of all known coelacanths, listing their place, date, depth, method and time of capture, their vital statistics and present whereabouts. For a complete listing of all coelacanths caught to date, contact the Coelacanth Conservation Council Secretariat, MTN ScienCentre, 1 Waterford Place, Century City, Milnerton 7441, Cape Town, South Africa.

Of the 175 specimens that have been recorded, the vast majority are lodged in museums and aquaria around the world, some of them on permanent display to the public. The following is a list of the places where it should be possible to view a coelacanth, although it would be wise to telephone in advance to check that it is on current display.

United States of America

American Museum of Natural History, New York City
Arizona State Museum of Natural History, Tuscon, Arizona
Field Museum of Natural History, Chicago
Los Angeles County Museum of Natural History, Los Angeles
Marine Vertebrate Collection, Scripps Institute of Oceanography,
    La Jolla
Museum of Comparative Zoology, Harvard University, Cambridge
Peabody Museum, Yale University, New Haven
Steinhart Aquarium, California Academy of Sciences, San Francisco

Canada

Institute of Ichthyology, Guelph University, Guelph
MacMillan Tropical Gallery, Public Aquarium of Vancouver
National Museum of Canada, Ottawa
Royal Ontario Museum, Toronto

United Kingdom

British Museum (Natural History), London
Cambridge Museum, Cambridge
Royal Scottish Museum, Edinburgh
The Royal Society, London
Ulster Museum, Belfast

France

Aquarium, Le Croisic
Château de la Bubrese
Hôtel Dieu, Lyon
Musée de la Pêche, Concarneau
Musée Zoologique de Strasbourg

Muséum National d'Histoire Naturelle, Paris
Museum of the Reunion, Paris
Museum of Oceanography, Quimper
National History Musem, Nantes
Natural History Museum, La Rochelle

Germany

Senckenberg Museum, Frankfurt
Staatliches Museum für Naturkunde, Stuttgart
Zoologische Staatssammlung, Munich

Denmark

University Zoological Museum, Copenhagen

Italy

Institute of Paleontology, Pavie
Zoological Museum, Turin

Sweden

Swedish Museum of Natural History, Stockholm

Netherlands

Natural History Museum, Leiden

Switzerland

Museum of Natural History, Geneva

Austria

Haus de Natur, Salzburg
Natural History Museum, Vienna

Belgium

Royal Institute of Natural Sciences, Brussels

South Africa

East London Museum
J.L.B. Smith Institute of Ichthyology, Grahamstown
Natal Museum, Durban
Transvaal Museum, Pretoria

Comoros

Centre National de Documentation et de la Réchèrche Scientifiques, Moroni
Island Ventures boathouse, Le Galawa Beach Hotel, Mitsamiouli, GC
Presidential Palace, Moroni
Sidi Bacari's Taxidermy workshop, near Mutsamudu, Anjouan

Madagascar

Institut Halieutique et des Sciences Marines, University of Toliara
National Museum, Antananarivo, Madagascar

Mozambique

Museum of Natural History, Maputo

Algeria

Centre de Recherches Océanographiques et de Pêche, Algiers

Zimbabwe

Queen Victoria Museum, Harare

Australia

Australian Museum, Sydney

Japan

Kanazawa Aquarium, Tokyo
National Science Museum, Tokyo
Yomiuri-Land Aquarium

China

Museum of Natural History, Beijing
Shanghai Museum
Specimen House, Institute of Vertebrate Paleontology and Paleoan-
thropology, Academia Sinica, Beijing
Specimen House of Ichthyology, Institute of Hydrobiology, Aca-
demica Sinica, Wuhan

Russia

Shirshov Institute of Oceanology, Moscow

Kuwait

Science and Natural History Museum

For a complete inventory of all coelacanths caught to date, contact the Coelacanth Conservation Council Secretariat, MTN ScienCentre, 1 Waterford Place, Century City, Milnerton 7441, Cape Town, South Africa

SELECTED READING

I am indebted to a large number of books and articles, only some of which I am able to acknowledge:

Books

Anthony, Jean: *Operation Coelacanthe* (Arthaud, 1976)

Barnett, Peter: *Sea Safari with Professor Smith* (S.A. Assoc. for Marine Biological Research)

Beebe, William: *Half Mile Down* (Harcourt Brace, 1934)

Broad, William: *The Universe Below* (Simon & Schuster, 1997)

Forey, Peter: *History of the Coelacanth Fishes* (Chapman & Hall, 1998)

Forte, Richard: *Life: An Unauthorised Biography* (HarperCollins, 1997)

Ley, Willy: *Exotic Zoology* (Viking Press, 1959).

Long, John A.: *The Rise of Fishes—500 Million Years of Evolution* (UNSW Press)

Millot, Jacques, and Anthony, Jean: *L'anatomie de* Latimeria chalumnae (Centre Nat. Res. Sci., 1960–78)

Millot, Jacques: *Le Troisième Coelacanthe* (Le Naturaliste Malgache, 1955)

Smith, J.L.B.: *Old Fourlegs: The Story of the Coelacanth* (Longman, Green, 1956)

Smith, J.L.B.: *Sea Fishes of Southern Africa* (Central News Agency, 1949)

Thomson, Keith: *Living Fossil* (Norton, 1991)

Ward, Peter Douglas: *On Methuselah's Trail* (W.H. Freeman, 1991)

Pamphlets

Greenwood, P.H.: *Latimeria chalumnae: The Living Coelacanth* (Ichthos pamphlet, 1993)

J.L.B. Smith Institute of Ichthyology: *The Life and Work of Margaret M. Smith* (N.D.)

J.L.B. Smith Institute of Ichthyology: *Ichthos: Tribute to Margaret Smith* (1987)

J.L.B. Smith Institute of Ichthyology: *Ichthos: The Coelacanth Jubilee* (1988)

J.L.B. Smith Institute of Ichthyology: *Ichthos: J.L.B. Smith Commemorative Edition* (1997)

Smith, Margaret M.: *J.L.B. Smith: His Life, Work, Bibliography and List of New Species* (Rhodes University, 1969)

Articles

Balon, E., Bruton, M., and Fricke, H.: "A Fiftieth Anniversary Reflection of the Living Coelacanth" (*Environmental Biology of Fishes*, 1988)

Bergh, W., Smith, W., Botha, W. and Laing, M.: "The Place of Natal Command in the History of World Science" (*Spectrum*, 1988)

Bruton, M.: "The Living Coelacanth Fifty Years Later" (*Transactions of the Royal Society of South Africa*, 1989)

Bruton, M.: "The Coelacanth—Can We Save It from Extinction?" (*World Wildlife Fund Reports*, 1989)

Bruton, M.: "The Mingled Destinies of Coelacanths and Men" (*Ichthos*, 1992)

Bruton, M., Cabral, Q., and Fricke, H.: "First Capture of a Coelacanth off Mozambique" (*S.A. Journal of Science*, 1992)

Conant, E.B.: "An Historical Overview of the Literature of Dipnoi" (*Journal of Morphology*, 1986)

Courtenay-Latimer, Eric: Diaries (unpublished, courtesy of Dr. M. Courtenay-Latimer)

Courtenay-Latimer, M.: "My Story of the First Coelacanth" (*Occidental Papers of the California Academy of Science*, 1979)

Courtenay-Latimer, M.: "Reminiscences of the Discovery of the Coelacanth" (*Cryptozoology*, 1989)

De Silva, D.: "Mystery of the Silver Coelacanth" (*Sea Frontiers*, 1966)

Dugan, J.: "The Fish" (*Colliers*, 1955)

Erdmann, M., Caldwell, R. and Moosa, K.: "An Indonesian 'King of the Sea'" (*Nature*, 1998)

Forey, P.: "Golden Jubilee for the Coelacanth" (*Nature*, 1988)

Forey, P.: "Blood Lines of the Coelacanth" (*Nature*, 1991)

Fricke, H.: "Im Reich der Lebenden Fossilien" (*Geo*, 1987)

Fricke, H.: "The Fish that Time Forgot" (*National Geographic*, 1988)

Fricke, H.: "Living Coelacanth: Values, Eco-ethics and Human Responsibility" (*Marine Ecology Progress Series*, 1997)

Fricke, H., and Hissmann, K.: "Natural Habitat of the Coelacanth" (*Nature*, 1990)

Fricke, H. and Plante, R.: "Habitat Requirements of the Living Coelacanth" (*Naturwissenschaften*, 1988)

Greenwood, P.H.: "Fifty Years a 'Living Fossil'" (*Biologist*, 1989)

Hall, M.: "The Survivor" (*Harvard Magazine*, 1989)

Heemstra, P., and Compagno, L.: "Uterine Cannibalism and Placental Viviparity in the Coelacanth? A Skeptical View" (*South African Journal of Science*, 1989)

Heemstra, P., et al.: "First Authentic Capture of a Coelacanth off Madagascar" (*South African Journal of Science*, 1996)

Hissmann, K., Fricke, H., and Schauer, J.: "Population Monitoring of the Coelacanth" (*Conservation Biology*, 1998)

Hissmann, K., and Schauer, J.: "Fossil Hunt" (*Diver*, 1991)

Millot, J.: "Notre Coelacanthe" (*Revue Madagascar*, 1953)

Millot, J.: "First Observations on a Living Coelacanth" (*Nature*, 1955)

Morris, A., and Morris, E.: "In Pursuit of the Coelacanth" (*Pacific Discovery*, 1973)

Munnion, C.: "Remembering Old Fourlegs" (*Optima*, 1988)

Plante, R., Fricke, H., and Hissmann, K.: "Coelacanth Population, Conservation and Fishery Activity at Grande Comore" (*Marine Ecology Progress Series*, 1998)

Schauer, J.: "The Privacy of a Living Fossil" (*Underwater*, 1992)

Smith, J.L.B.: "A Living Fish of the Mesozoic Type" (*Nature*, 1939)

Smith, J.L.B.: "A Surviving Fish of the Order Actinistia" (*Transactions of the Royal Society of South Africa*, 1939)

Smith, J.L.B.: "A Living Coelacanth Fish from South Africa" (*Transactions of the Royal Society of South Africa*, 1940)

Smith, J.L.B.: "The Second Coelacanth" (*Nature*, 1953)

Smith, M.: "The Search for the World's Oldest Fish" (*Oceans*, 1970)

Stobbs, R.: "The Coelacanth Enigma" (*The Phoenix*, 1989)

Stobbs, R.: "The Comoro Islands Traditional Artisanal Fishery" (*Ichthos*, 1990)

Stobbs, R.: "Eric Ernest Hunt—The Aquarist" (*Ichthos*, 1996)

Stobbs, R.: "Hiraiako—The Broken Thread" (*Ichthos*, 1996)

Stobbs, R.: "Gone Fishin'—for a Purgative" (*Ichthos*, 1998)

Vicente, N.: "Un Coelacanth à Madagascar" (*Oceanorama*, 1997)

White, E.I.: "One of the Most Amazing Events in the Realm of Natural History in the Twentieth Century" (*London Illustrated News*, 1939)

INDEX

Note: Numbers in *italics* refer to illustrations.